Ternary Equilibrium Diagrams

Ternary Equilibrium Diagrams

D. R. F. WEST
B.Sc, Ph.D, D.I.C, C.Eng, F.I.M
Imperial College of Science and Technology
University of London

SECOND EDITION

LONDON NEW YORK
CHAPMAN AND HALL

First published 1965 by
Macmillan and Company Limited
St Martin's Street, London WC2

Second edition published 1982 by
Chapman and Hall Ltd
11 New Fetter Lane, London EC4P 4EE
Reprinted 1985
Published in the USA by
Chapman and Hall
733 Third Avenue, New York NY 10017

Printed in Great Britain at the
University Press, Cambridge

ISBN 0 412 22970 6 (cased)
ISBN 0 412 22820 3 (paperback)

Library of Congress Cataloging in Publication Data
West, D. R. F. (David Richard Frederick)

Ternary equilibrium diagrams.
Bibliography: p.
Includes index
1. Phase diagrams, Ternary. I. Title.
TN690.W468 1981 660.2′963 81-14060
ISBN 0-412-22970-6 AACR2
ISBN 0-412-22820-3 (pbk.)

British Library Cataloguing in Publication Data
West, D. R. F.

Ternary equilibrium diagrams.—2nd ed.
1. Ternary systems 2. Phase diagrams
I. Title
541.3′63 QA141.8.T45
ISBN 0-412-22970-6
ISBN 0-412-22820-3 Pbk

Contents

Preface to the First Edition

INFORMATION concerning phase equilibria, such as can be represented by equilibrium diagrams is important in a number of fields of scientific study, and especially in metallurgy, ceramics, and chemistry. Materials of interest in these fields range through single-component to multi-component systems, the latter often being very complex. While many industrially important systems can be represented adequately by binary equilibrium diagrams, ternary diagrams provide a basis for studying a wide range of the more complex systems, such as are encountered in certain industrial alloys, and in slags and ceramics.

A number of texts dealing with ternary systems are already available (see for example references 2–8). Some of these are very comprehensive and include many diagrams, and their use will take the student of the subject to an advanced and detailed level of understanding. The present monograph is intended primarily as an introductory text, which it is hoped will prove useful for undergraduate and postgraduate students of metallurgy and ceramics, in particular. Special attention is given to the requirement of the metallurgist and ceramist to use phase diagrams as a means of understanding phase changes that occur during heating and cooling, as in solidification and heat treatment of alloys. This aspect is emphasized in terms of the principles of solidification reactions in relation to solid state constitution, particularly making use of liquidus and solidus projections.

Hypothetical systems are considered, commencing with simple examples and then proceeding to those of greater complexity. The sequence is designed to show principles and is illustrated, in some cases, by reference to actual systems. The general treatment, in itself complete, nevertheless assumes a previous knowledge of binary phase equilibria.

The author wishes to thank Dr D. Lloyd Thomas for reading and commenting on the manuscript.

D. R. F. West

Preface to the Second Edition

THE second edition of this monograph retains the aim of the original text to provide an introductory treatment of ternary equilibrium diagrams. However, two significant modifications have been made to assist the reader in obtaining a detailed understanding of the subject and to emphasize the practical importance of ternary diagrams. These modifications are the provision of problems with outline solutions and answers and the addition of a substantial chapter to extend the coverage of actual systems.

The actual ternary systems, presented as 'case studies' in this chapter, are of practical relevance in various fields of metallurgy and materials science such as extraction and process metallurgy, industrial alloys, ceramics and semiconductors. Some of these systems show considerably greater complexity than those considered in the previous chapters, and they include examples of solid state transformations. The general approach in the new chapter is to outline the main features of the systems and to relate them to practical applications. It is hoped that the sequence chosen for the presentation of the examples will be helpful in the study of complex systems.

The author wishes to acknowledge gratefully the valuable advice and assistance provided by Mr Alan Prince, Dr D. Lloyd Thomas, Professor M. C. Flemings and Dr F. P. Glasser. Acknowledgement is made to the University of London for permission to reproduce examination questions.

D. R. F. West
January 1981

1
General Considerations

The phase rule

THE Phase Rule as propounded by Willard Gibbs[1] underlies the study of phase equilibria and is of basic importance in the representation of such equilibria by means of equilibrium (phase) diagrams. Reference should be made to other sources[3, 6] for accounts of the derivation of this Rule and the assumptions involved, and here only a brief indication of its significance is given.

A particular state of equilibrium is characterized by the number and identity of the phases present, and the Phase Rule expresses the relationship, at equilibrium, of the number of phases, P, with the number of components, C, and the number of degrees of freedom, F. The rule may be expressed as follows:

$$P + F = C + 2$$

The *phases* are the homogeneous parts of a system, bounded by surfaces and, in principle at least, can be mechanically separated from one another. The term phase is often used less precisely in practice to include portions of a system that are not entirely homogeneous, and are hence not strictly in equilibrium, e.g. phases in solid alloys often show compositional variations.

The *number of components* is the smallest number of independently variable constituents necessary for the statement of the compositions of all the phases in the system; in alloy systems, for example, the number of constituent metals.

The number of degrees of freedom of the equilibrium state is the number of conditions that can be altered independently without changing the state of the system, or which have to be specified to define completely the state of the system. In the above statement of the Phase Rule it is assumed that temperature, pressure and composition are the only externally controllable variables that influence the phase equilibria. In ternary systems there are two compositional variables to consider, since a statement of the concentrations of two components is sufficient to define the composition, e.g. in an alloy of metals A, B, and C, the specification of the percentages of, say, A and B suffices to establish the alloy composition.

In many instances in metallic systems, the vapour pressures of the liquid and solid phases are negligible, or small, in comparison with atmospheric pressure; the pressure may then be considered as a constant, and the Phase Rule reduces to

$$P + F = C + 1$$

One degree of freedom is exercised by fixing the pressure in this way. Applying this 'reduced rule' to a ternary system ($C = 3$), the general position is summarized as follows:

P	1	2	3	4
F	3	2	1	0

The present account considers only those systems in which the pressure variable can be neglected.

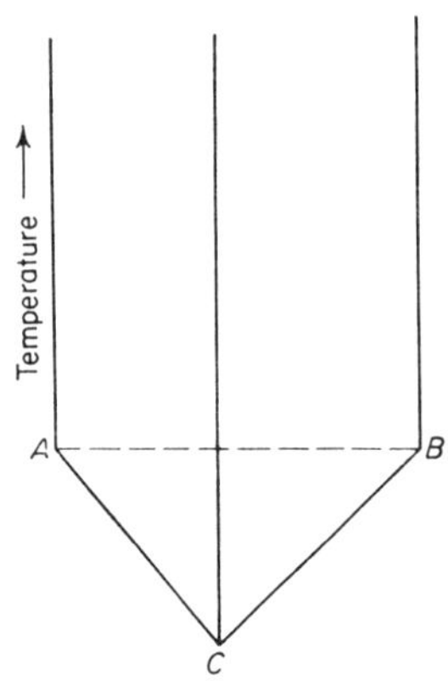

Fig. 1.1. Space model used for representing ternary systems

The ternary space model

To represent completely the phase equilibria at constant pressure in a ternary system, a three-dimensional model, commonly termed a space model, is required; the representation of composition requires two dimensions, and that of temperature, a third dimension. The model used is a triangular prism (Fig. 1.1), in which the temperature is plotted on the vertical axis, and the composition is represented on the base of the prism, which may be conveniently taken as an equilateral triangle. Thus, in Fig. 1.1, the vertical sides of the prism represent the three binary systems AB, BC, and AC that make up the ternary system ABC.

Regarding the composition triangle the three corners represent the three components, points on the sides represent binary alloys, and points within the triangle, ternary alloys. Generally, the length of each side of the triangle is considered as divided into 100 equal parts each of which represents 1% on the composition scale in each of the binary systems.

Composition scales are commonly plotted in weight percentages. Then, the percentage by weight of each component in a ternary alloy (e.g. alloy x in Fig. 1.2) is obtained by the following construction: a line mn is drawn parallel to side BC and the percentage of A is then given by the lengths mB or nC. The percentage of B is given by the lengths oA or pC, obtained

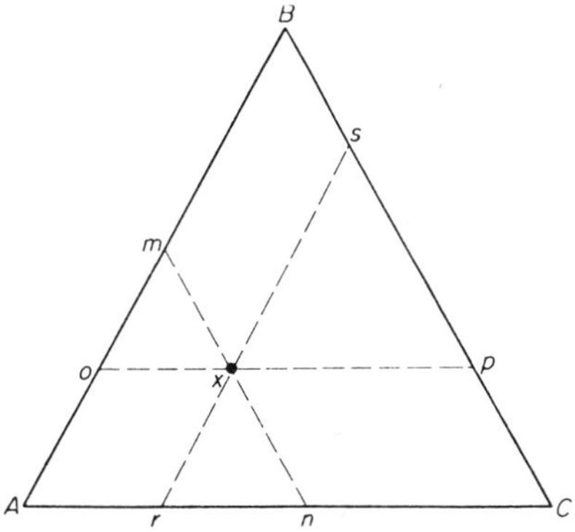

Fig. 1.2. Equilateral triangle to represent compositions

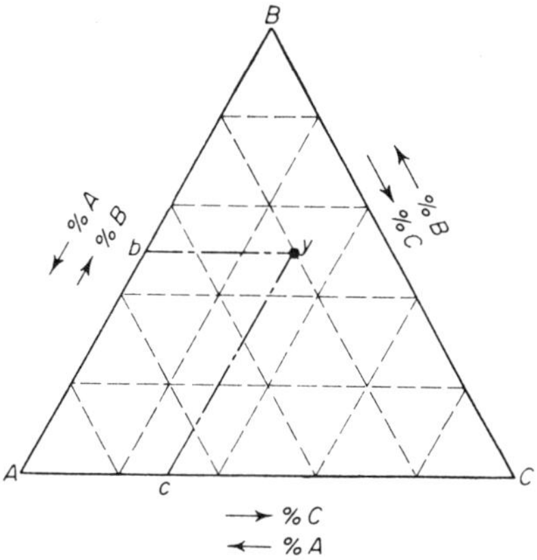

Fig. 1.3. Plotting alloy compositions on the compositional triangle

by drawing line op parallel to side AC. Similarly, the percentage of C is obtained by drawing rs parallel to side AB and determining the lengths rA or sB.

Thus

$$\% \text{ A} = mB = nC = xs = xp$$
$$\% \text{ B} = oA = pC = xr = xn$$
$$\% \text{ C} = rA = sB = xo = xm$$

To facilitate the plotting of composition points, the equilateral triangle may be ruled with lines parallel to its sides (Fig. 1.3).

Consider the plotting of the composition point of an alloy y, containing 50% B and 30% C. Starting from point A, a length Ab corresponding to 50% B is marked out along AB, and then a length Ac corresponding to 30% C is marked out along AC. The composition point of the alloy then lies at the intersection of the appropriate ruled lines as shown.

An alternative, but less convenient method of representing the composition on the equilateral triangle is to take the altitude of the triangle as 100%, in which case the percentages of the individual components in a ternary alloy are given by the heights of the respective perpendiculars from the composition point to the opposite sides of the triangle.

A right-angled triangle is sometimes used to represent compositions. In the example shown in Fig. 1.4 the percentage of A is obtained by subtracting the percentages of B + C from 100. This method of representation is useful

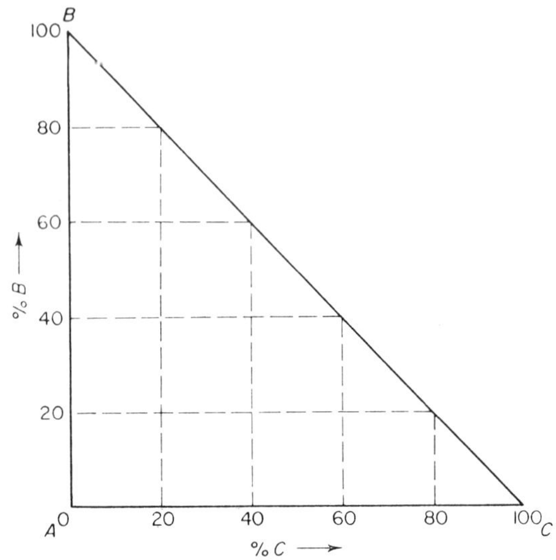

Fig. 1.4. Right-angled triangle to represent compositions

when interest is centred on alloys of metal A, containing relatively small amounts of B and C, that is when the representation of the A-rich corner of the system is required, as is often the case in aluminium-base alloys.

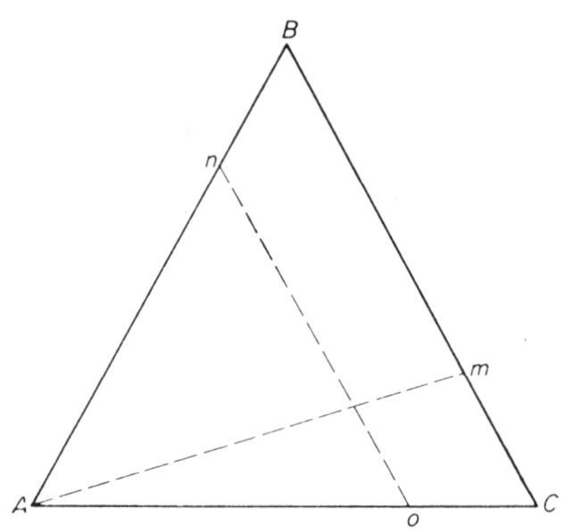

Fig. 1.5

The following features are of interest in connection with compositional aspects of ternary systems.

(1) If a straight line such as *Am* in Fig. 1.5 is drawn from one corner

of the triangle (i.e. from component A) to intersect the opposite side (i.e. *BC*), then the ratio of the other two components (i.e. B and C) is constant in alloys represented by points lying on the line. Considering alloy compositions along *Am*, as the composition point becomes progressively further away from *A* the percentage of A present in the alloys decreases, but the relative amounts of B and C remain the same.

(2) A line drawn parallel to one side of the triangle is a line along which alloys contain a constant percentage of one component, i.e. along *no* (Fig. 1.5) the percentage of A remains the same.

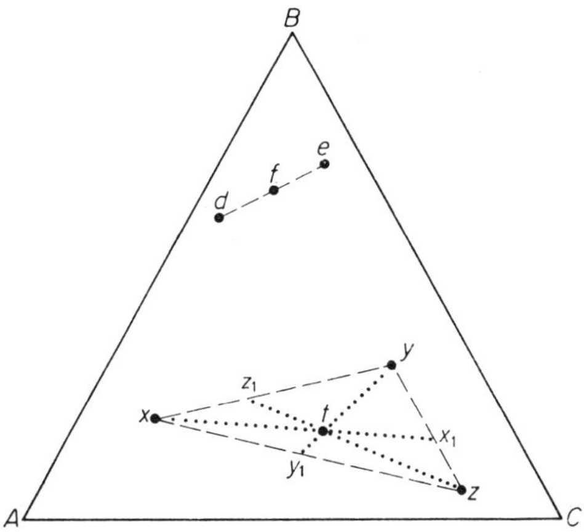

Fig. 1.6. Mixtures of two or three alloys or phases

(3) If two alloys (e.g. d and e in Fig. 1.6) are mixed together, the composition *f* of the resulting mixture lies on a straight line joining the compositions of the two alloys. The total composition lies at the point dividing the line into segments inversely proportional to the ratio of the quantities of the original alloys. Then, using weight percentages:

$$\frac{\text{Weight of d}}{\text{Weight of e}} = \frac{ef}{df}$$

$$\% \text{ d} = \frac{ef}{de} \times 100$$

$$\% \text{ e} = \frac{df}{de} \times 100$$

This is the so-called 'lever principle' or 'lever rule' which finds an important application in 'tie-lines' when an alloy decomposes into a two-phase mixture (as discussed later).

(4) If three alloys x, y, and z are mixed together, the composition of the mixture, t (Fig. 1.6), lies within the triangle formed by joining the compositions of the alloys by straight lines. Using the 'centre of gravity' principle, the triangle may be considered as being supported on a point fulcrum at t and having masses at its corners proportional to the amounts of the alloys. Then, in equilibrium, the position of t is such that the percentages of the three alloys in the mixture are given by

$$\% \text{ x} = \frac{x_1t}{xx_1} \times 100$$

$$\% \text{ y} = \frac{y_1t}{yy_1} \times 100$$

$$\% \text{ z} = \frac{z_1t}{zz_1} \times 100$$

The application of this principle in 'tie-triangles', representing three-phases in equilibrium, will be discussed later.

In general, except for relatively simple cases, it is difficult to depict clearly by means of perspective diagrams the complete ternary space model with its constituent spaces, curves, and points. It is convenient to use projected views (for example, a projection of the liquidus or the solidus surface on to the base of the space model), and also plane sections taken either horizontally (i.e. isothermally) or vertically through the space model. These types of diagrams are very useful when considering the solidification processes of alloys, or solid state transformations relevant to heat treatments, and are widely used for the presentation of experimental data for ternary systems.

Phases and their equilibria

The present account deals with equilibria between liquid and solid phases in ternary systems. In the liquid state the components of many systems are completely soluble in one another in all proportions and therefore form a complete range of solutions. In cases where the components are not soluble in one another in all proportions there exists a region, or regions, in which two or three liquids co-exist; such a region can be termed a 'miscibility' or 'solubility gap'.

In the solid state the phases are either the pure components of the system, solid solutions, or intermediate phases. In a substitutional solid solution each type of atom involved occupies positions on a common lattice, characteristic of the solvent element, while an interstitial solution has the solute atoms located in the interstices of the lattice of the solvent.

In a ternary solution, one type of atom may be located substitutionally and another interstitially. Complete solid solubility is shown in some cases, but partial solid solubility is more common.

Many systems contain one or more intermediate phases that occur over certain ranges of composition and temperature; such phases generally differ from the components with respect to lattice structure. Some intermediate phases are stable up to their melting point, while others form from the liquid by a peritectic reaction; in alloy systems several types may be distinguished on the basis of atomic bonding and compositional ranges. In some ternary systems the only intermediate phases present may be those derived from one or more of the binary systems, but there are many cases where true ternary phases occur, i.e. those whose composition ranges do not extend into the binary systems. When several intermediate phases occur the phase diagrams may be complex because of the large number of possibilities for the equilibria between the phases. In particular, many ceramic systems show very great complexity in this respect.

It is relevant to consider briefly, in the light of the Phase Rule, the representation of equilibria involving various numbers of phases. For this purpose it is assumed that the solid phases are of variable composition, and comparison is made of equilibria in binary and ternary systems, making reference to a hypothetical binary eutectic system showing partial solid solubility (Fig. 1.7). It should be noted that constant pressure is assumed and the reduced Phase Rule is used.

One phase is present

In the binary system, when only one phase is present there are two degrees of freedom, and to define the state of the system both the temperature and the concentration of one component must be specified; in the binary alloy diagram an area represents the range of existence of the phase (e.g. the liquid phase region in Fig. 1.7). In a ternary system, application of the Phase Rule shows that there are three degrees of freedom; thus it is necessary to define the temperature and the concentrations of two components to define the state. The phase region is represented by a three-dimensional space.

Two phases co-exist

When two phases co-exist in equilibrium in the binary system, as for example in the $L + \beta$ region of Fig. 1.7, there is one degree of freedom. The two-phase region may be viewed as being made up of a stack of tie-lines (such as $L_1\beta_1$); each of these isothermal lines joins the composition points of the two phases in equilibrium at a particular temperature. The ends of the tie-lines at the various temperatures constitute the liquidus and solidus curves respectively. Thus, the composition points for each of the phases form a curve and the equilibrium is defined if one variable is stated, i.e. either the temperature, or the concentration of a component in one of the phases.

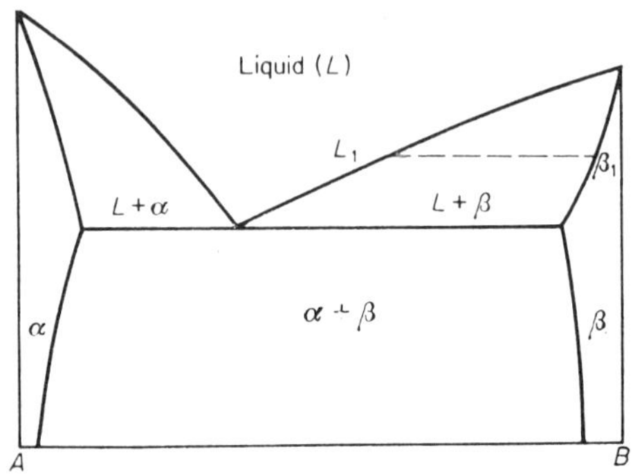

Fig. 1.7. Binary eutectic system showing partial solid solubility

In a ternary system, when two phases co-exist, there are two degrees of freedom and two variables must be defined, e.g. the temperature, and a composition term for one of the phases. The two-phase region in the diagram is a space that may be considered as being filled with tie-lines whose extremities constitute the two surfaces bounding the space. Thus, the composition of each of the phases in equilibrium lies on one of these surfaces, not on a curve as in the binary system; this is a consequence of the extra degree of freedom in the ternary system as compared with the binary.

Three phases co-exist

In the binary system, when three phases co-exist in equilibrium, there are no degrees of freedom, i.e. the system is invariant. Thus, at the eutectic temperature in Fig. 1.7, the compositions of each of the phases, liquid, α and β, respectively, are represented by fixed points and the equilibrium is depicted in the diagram by an isothermal line.

In a ternary system, one degree of freedom remains when three phases co-exist, and the state is defined if either the temperature or a composition term for one of the phases is specified. The three-phase region in the ternary diagram is a space bounded by three surfaces and three curves formed by the intersection of pairs of these surfaces. An isothermal section through this space is a tie-triangle, as shown for example in Fig. 1.8 for equilibrium between a liquid, L, and two solid solutions α and β. The sides of such a triangle are tie-lines, and the three corners represent the compositions of the three co-existing phases.

The three-phase space can be considered as being made up of a stack of such tie-triangles; the positions, at various temperatures, of the tie-lines bounding the triangle, mark out the boundary surfaces ('ruled' surfaces) of the region and the corresponding corner positions of the triangle mark out the boundary curves. Thus, the compositions of each of the three phases are located along curves in the space diagram, as compared with their representation by points for the case of three-phase equilibrium in a binary system. Because of the additional degree of freedom in the case of a ternary system, three-phase reactions (e.g. of the eutectic or peritectic types) occur over a range of temperatures, and not at a fixed temperature as in a binary system.

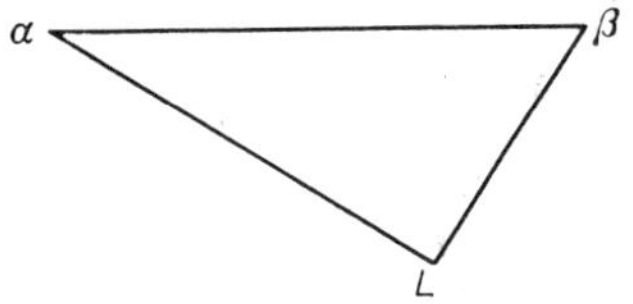

Fig. 1.8. Tie-triangle representing the equilibrium between liquid (L) and two solid solutions α and β

Four phases co-exist

With binary systems, the possible co-existence of four phases arises if pressure is taken into account as a variable, and if the complete diagram of temperature, pressure, and composition is considered.

In ternary systems, applying the reduced Phase Rule, the co-existence of four phases is a case of invariant equilibrium since there are no degrees of freedom. Such an equilibrium (e.g. of the ternary eutectic and peritectic types) exists only at a particular temperature in ternary systems, and the compositions of each of the four phases lie at fixed points on a plane in the space model corresponding to the particular temperature.

Classification of ternary systems

It is convenient to consider types of systems containing various numbers of phases, and on this basis the classification given below is adopted in the following pages; this classification also defines the types of phases and reactions dealt with in Chapters 2–5 respectively.

Chapter 2 Systems containing only two phases, namely a liquid + a solid solution.

Chapter 3 Systems containing three phases (e.g. a liquid + 2 primary solid solutions) and showing a three-phase solidification reaction.

Chapter 4 Systems containing four phases (e.g. a liquid + 3 primary solid solutions) and showing a four-phase invariant reaction.

Chapter 5 Systems containing more than four phases (e.g. a liquid + 3 primary solid solutions + one or more intermediate phases) and showing two or more invariant reactions.

In general, complete miscibility in the liquid state is assumed; however, a liquid miscibility gap, with an associated monotectic-type reaction is briefly discussed in Chapter 3. Consideration of phase transformations in these systems is based on the assumption that equilibrium conditions are attained.

Problem

In a system ABC, a ternary alloy of composition 30 wt% B and 30 wt% C consists at a particular temperature of three phases of equilibrium compositions as follows:

Liquid phase 50% A, 40% B, 10% C
α solid solution 85% A, 10% B, 5% C
γ solid solution 10% A, 20% B, 70% C

(a) Calculate the proportions by weight of liquid, α and γ present in this alloy.

(b) For the same temperature deduce the composition of the alloy which will consist of equal proportions of α and γ phases of the compositions stated above, but with no liquid phase.

2

Systems Containing Two Phases

THE system depicted in the view of the space model shown in Fig. 2.1 consists of three components A, B, and C that are completely soluble in one another in both the liquid and the solid state. The liquidus and the solidus, which are represented by curved surfaces, enclose a space constituting the two-phase region of liquid + solid solution. This two-phase region may be considered as consisting of a bundle of tie-lines of varying direction; each tie-line links the composition points of liquid and solid co-existing at a

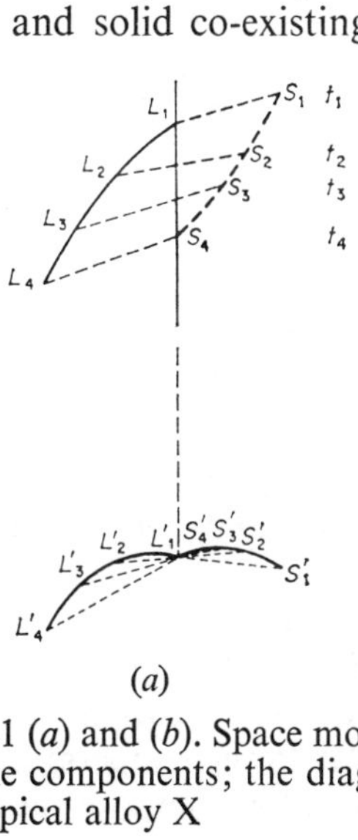

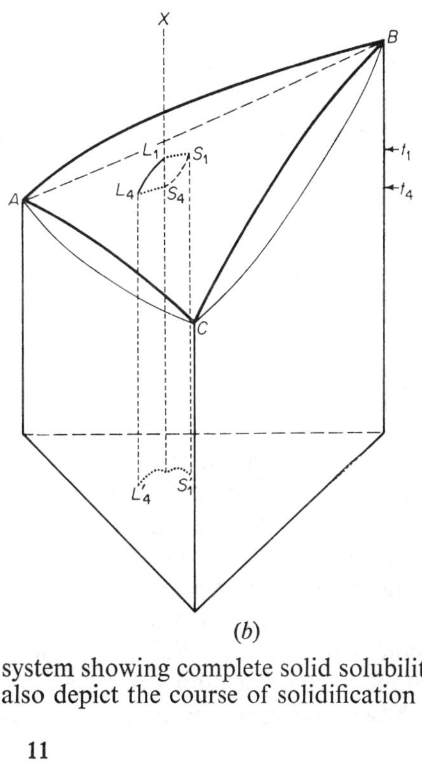

(*a*) (*b*)

Fig. 2.1 (*a*) and (*b*). Space model of system showing complete solid solubility of the components; the diagrams also depict the course of solidification of a typical alloy X

particular temperature, and no two tie-lines at the same temperature can cross one another. Experiment is required to establish the position and the direction of the tie-lines in any particular system, and relatively little data of this type are available for actual systems.

The experimental determination of liquidus and solidus temperatures for a series ot alloys ın a system provides data for plotting isotherms on the liquidus and solidus surfaces. Such isotherms may be conveniently depicted on a projected view on the base of the space model, e.g. Fig. 2.2 show a liquidus projection or 'liquidus plan' for the Ag–Au–Pd system.[9]

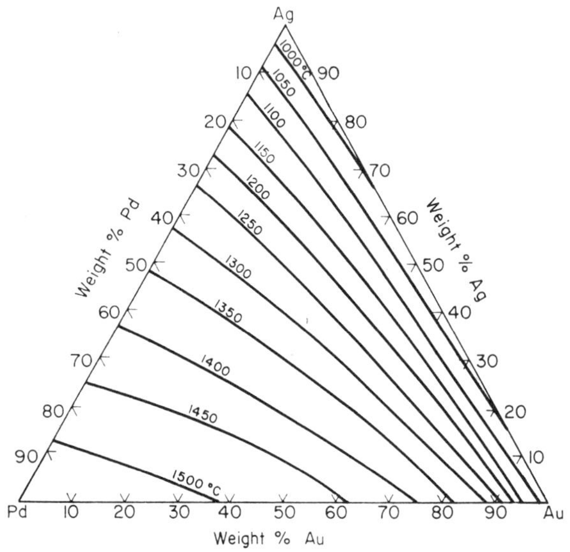

Fig. 2.2. Liquidus projection for the Ag–Au–Pd system (From *Metals Handbook*, 7th Edn., 1948)

It should be noted that in some systems the liquidus and solidus surfaces may exhibit a maximum or a minimum point at which the liquidus and solidus touch one another.

The process of equilibrium solidification in the system shown in Figs. 2.1*a* and *b* may be illustrated by reference to an alloy of composition indicated by point *X*. On cooling through the liquidus temperature to temperature t_1, solidification begins by the formation of crystals of solid solution which contain more of component B than the original liquid. The compositions of the liquid and solid are given by the extremities of the tie-line L_1S_1. The positions of several tie-lines at lower temperatures are also shown and each tie-line passes through the alloy composition point.

The extremities of these tie-lines lie on the liquidus and solidus surfaces respectively, and the paths $L_1L_2L_3L_4$ and $S_1S_2S_3S_4$ represent the progressive change of composition of liquid and solid during solidification. At any temperature application of the lever rule to the appropriate tie-line enables the relative amounts of liquid and solid to be calculated, e.g. at t_2 (Fig. 2.3*a*).

$$\frac{\text{Weight of liquid}}{\text{Weight of solid}} = \frac{X_2S_2}{X_2L_2}$$

$$\%\ \text{Liquid} = \frac{X_2S_2}{L_2S_2} \times 100$$

$$\%\ \text{Solid} = \frac{X_2L_2}{L_2S_2} \times 100$$

With falling temperature, the percentage of liquid decreases until at the solidus temperature, t_4, the composition of the solid solution attains the alloy composition and no liquid remains. The case of non-equilibrium freezing, leading to coring, and the resultant extension of the temperature range of solidification has been described by Rhines.[7]

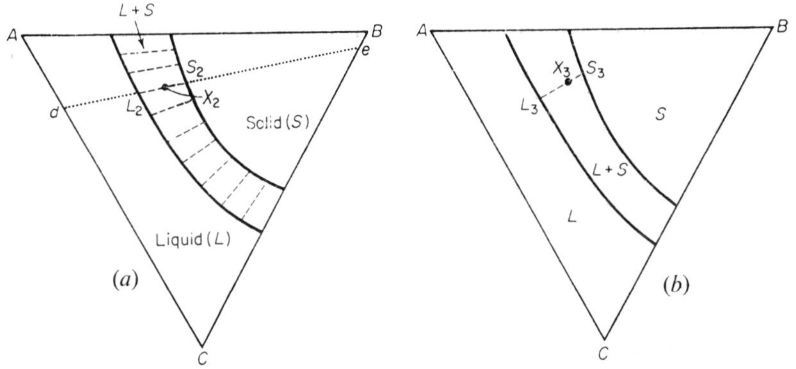

Fig. 2.3 (*a*) and (*b*). Isothermal sections through the space model of Fig. 2.1 for temperatures t_2 and t_3 respectively

As stated above, if it is required to know the actual compositions of liquid and solid co-existing at any temperature, experimental determination is necessary. For a given alloy, the tie-lines at various temperatures do not generally lie in the same vertical plane and the paths of the extremities of the tie-lines on the liquidus and solidus are generally curved. This effect also is illustrated in Figs. 2.1*a* and *b* by the projected view on the base of the space model of paths representing the change in composition of liquid and solid. The tie-line L_1S_1 represents the beginning of solidification, and L_4S_4 the end of solidification.

Tie-lines for alloy X are also indicated in the isothermal sections through the space model at temperatures t_2 and t_3 respectively (Figs. 2.3*a* and *b*). Other typical tie-lines for the liquid + solid region are included in the sections, and it is only by coincidence that a tie-line may point directly towards a corner of the system.

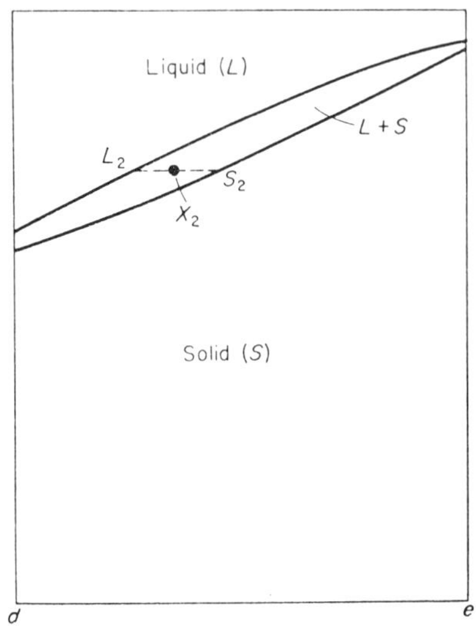

Fig. 2.4. Vertical section taken along the direction of the tie-line L_2S_2 of Fig. 2.3 (*a*)

The vertical section *de* shown in Fig. 2.4 is taken along the direction of the tie-line L_2S_2 at temperature t_2. At other temperatures within the liquidus–solidus range, the tie-lines intersect the plane of the section at the composition point of the alloy, but the extremities of the tie-lines do not lie in the plane. Thus a vertical section of this type through a ternary system differs essentially from a binary phase diagram since, in general, the tie-lines do not lie in the plane of the section. During the solidification of alloys whose compositions lie in the section, the liquid and solid compositions change out of the plane of the section. On such a section it is not therefore correct to assume that a horizontal line cutting the liquidus and solidus surfaces in this plane necessarily defines the compositions of the liquid and solid co-existing in equilibrium.

Problem

The melting points of A, B and C are 1000°C, 900°C and 750°C, respectively. The constituent binary systems of the ternary system ABC each show complete solubility in both the liquid and the solid states. The following data refer to three binary alloys:

Composition (wt %)			°C	
A	B	C	Liquidus	Solidus
50	50	—	975	950
50	—	50	920	850
—	50	50	840	800

For the ternary system draw and label diagrams consistent with the above data to represent the possible forms of (a) a liquidus projection, showing isotherms for 950°C and 850°C, (b) a solidus projection showing isotherms for 950°C and 850°C, (c) an isothermal section at 950°C, (d) a vertical section joining *A* to the mid-point of *BC*. Assume that the liquidus and solidus surfaces do not exhibit a maximum or a minimum.

3
Systems Containing Three Phases

THE following cases will be considered:

(*a*) Liquid(L) + two primary solid solutions (α and β) giving a three-phase eutectic reaction $L \longrightarrow \alpha + \beta$;
(*b*) Liquid(L) + two primary solid solutions (α and β) giving a three-phase peritectic reaction $L + \alpha \longrightarrow \beta$;
(*c*) Transition between eutectic and peritectic reactions;
(*d*) Two liquids (L_1 and L_2) + one primary solid solution (α) giving a three-phase monotectic reaction

$$L_1 \longrightarrow L_2 + \alpha$$

(*e*) Systems containing a solid state miscibility gap that closes within the system.

Systems containing a eutectic reaction $L \longrightarrow \alpha + \beta$

In the hypothetical system ABC shown in Fig. 3.1 it is assumed that there is complete liquid miscibility. Binary system AC shows complete solid solubility, while systems AB and BC show partial solubility and are of the eutectic type. The solid solutions are termed α and β respectively, the former extending completely across the AC side of the system; in the ternary system the single-phase regions of α and β are separated from each other by a two-phase $\alpha + \beta$ region. The temperature of the eutectic point M representing the reaction $L \longrightarrow \alpha + \beta$ in system AB is assumed to be higher than that of point N representing the reaction $L \longrightarrow \alpha + \beta$ in system BC.

The view of the space model (Fig. 3.1) shows that the liquidus is made up of two surfaces $AMNC$ and BMN, corresponding to the primary solidification of α and β respectively. These surfaces intersect along a eutectic curve or 'valley' *MN*. In principle, this curve *MN* may show a maximum or minimum point, but it is assumed in the example illustrated that the curve falls smoothly as it traverses the system from side *AB* to side *BC*. *DG* and *EF* are curves joining the points representing the respective compositions of the α and β phases formed in the eutectic reactions in the binary systems (these reactions are represented by the horizontal straight lines *DME* and *GNF*). Surfaces *ADGC* and *BEF* form part of the solidus of the system, and each of the liquid + solid regions (i.e. liquid + α and liquid + β) enclosed by the respective liquidus and solidus surfaces resembles the two-phase region of the solid solution system previously discussed.

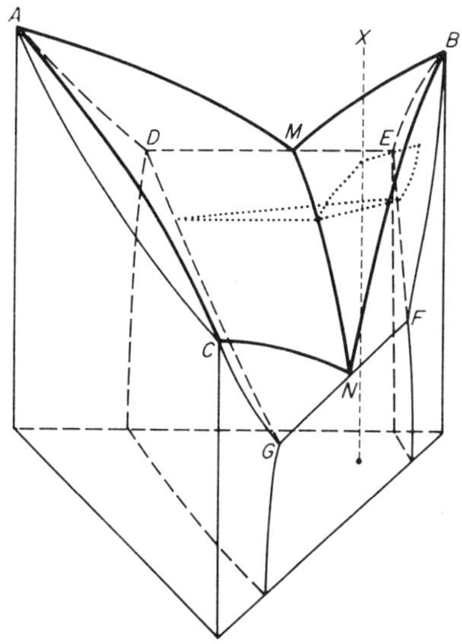

Fig. 3.1. Space model of system showing a eutectic reaction $L \longrightarrow \alpha + \beta$

The three curves *MN*, *DG*, and *EF* do not lie in the same plane; curve *MN* lies above the surface *DEFG*, so that there are three curved surfaces, namely *DMNG*, *MEFN*, and *DEFG*, which enclose a three-phase space, representing the co-existence of liquid, α and β. Each of these surfaces is made up of tie-lines, representing the co-existence of *L* and α, *L* and β, and α and β respectively. Surfaces *DMNG* and *MEFN* separate the three-phase space from the $L + \alpha$ and $L + \beta$ regions respectively: *DEFG* separates the three-phase space from the $\alpha + \beta$ region, and is thus the solidus surface for the $\alpha + \beta$ portion of the system. (The form of a vertical section that cuts through the three-phase space is shown in Fig. 3.8.) Where the three-phase space terminates in the binary systems *AB* and *BC* it shrinks to the horizontal straight lines *DME* and *GNF* respectively. [In a later section (page 36) reference will be made to a ternary system in which the three-phase space closes within the system.]

The eutectic reaction

The three-phase space is associated with the eutectic reaction $L \longrightarrow \alpha + \beta$ which occurs over a range of temperature in a ternary system, not at a constant temperature as in a binary system. The three-phase region may be considered as being made up of a stack of isothermal tie-triangles, and its form is illustrated in Fig. 3.2 which shows typical tie-triangles at several temperatures. The corners of these triangles represent the compositions of the phases co-existing in equilibrium at the particular temperatures concerned, and as the eutectic reaction proceeds with falling temperature the liquid composition changes along curve MN, and the compositions of α and β change along DG and EF respectively. Thus, while three phases co-exist, the composition of each phase is restricted to a curve as the temperature changes.

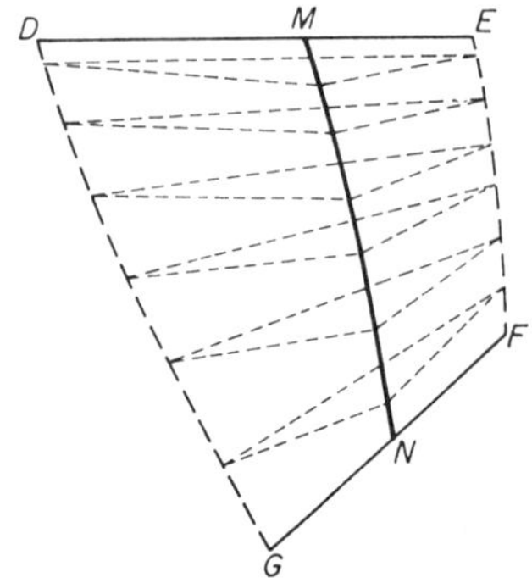

Fig. 3.2. Three-phase region (liquid, α and β) of Fig. 3.1

The process of eutectic solidification under equilibrium conditions may be illustrated by reference to the alloy X shown in Fig. 3.1; this alloy, when solid, consists of primary crystals of β plus a eutectic mixture of $\alpha + \beta$. During solidification, the primary stage (i.e. the separation of β), is the same in principle as the solidification of an alloy in a system showing complete solid solubility. Thus, the composition of the liquid changes along a path on the liquidus surface as the temperature falls, and the β-composition changes on a path on the solidus. Then at a certain temperature t_1, before solidification is completed, the liquid composition reaches a point on curve MN and simultaneously the β composition reaches a point on curve EF. The situation at t_1 is illustrated by the tie-triangle in Fig. 3.1 and in the projected view of the tie-triangle shown in Fig. 3.3a; the alloy composition point is seen to lie on the $L\beta$ tie-line that forms one of the boundaries of the $L\alpha\beta$ triangle at this temperature. The relative amounts of L and β may be obtained by applying the lever principle to the tie-line. With a fall in temperature to t_2, the three-phase region is entered (i.e. liquid, α and β co-exist), and the alloy composition point now lies inside the tie-triangle (Fig. 3.3b). The compositions of the liquid, α and β are given by the points, L_2, α_2, and β_2 respectively, which are the corner points of the triangle, and application of the 'centre-of-gravity' principle gives the amounts of the phases present.

$$\% \text{ Liquid } (L_2) = \frac{XL_2'}{L_2L_2'} \times 100$$

$$\% \ \beta_2 = \frac{X\beta_2'}{\beta_2\beta_2'} \times 100$$

$$\% \ \alpha_2 = \frac{X\alpha_2'}{\alpha_2\alpha_2'} \times 100$$

In the present case it is assumed that the tie-triangles have approximately the same size and shape throughout the freezing range of the alloy X, but in actual systems there can be appreciable differences in these characteristics of the triangles at the various temperatures.

During cooling to some lower temperature, t_3, the compositions of liquid, α and β change along the curves *MN*, *DG*, and *EF* respectively. The tie-triangle corresponding to temperature t_3 is shown in Fig. 3.3*c*.

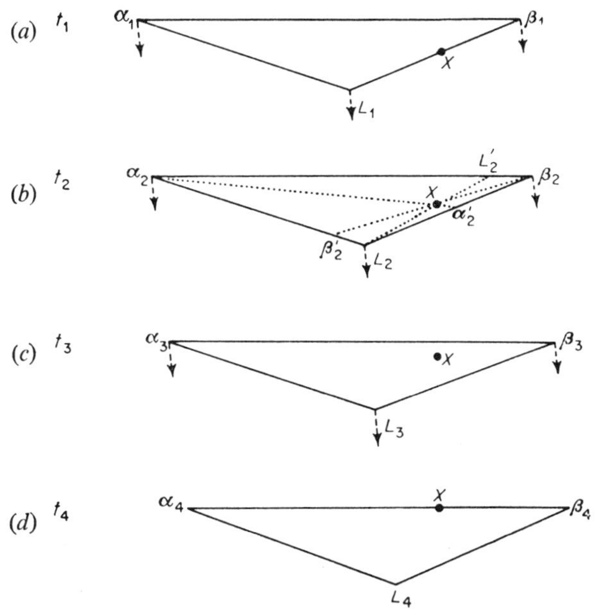

Fig. 3.3. Tie-triangles at various temperatures illustrating the progress of the eutectic reaction in alloy X of Fig. 3.1

The position of the triangle relative to the alloy composition point X is different from that at t_2, and application of the 'centre-of-gravity' principle shows that the amount of liquid is less at t_3 than at t_2, while the amounts of α and β are greater; this fact is in accordance with the eutectic nature of the reaction, $L \longrightarrow \alpha + \beta$.

Compositional changes in the liquid, α and β continue during further cooling until at the solidus temperature, t_4, the eutectic reaction is completed. The situation is depicted by the tie-triangle in Fig. 3.3*d*, where the alloy composition point is now seen to lie on the $\alpha + \beta$ side of the tie-triangle, that is to say, no liquid remains. The relative amounts of α and β may be obtained by applying the lever principle to the $\alpha\beta$ tie-line.

The process of eutectic solidification as described, may be viewed in terms of the displacement of the tie-triangles, relative to the alloy composition point, solidification being completed when the $\alpha\beta$ side of a triangle contains the alloy composition point.

Projected view of the system

The solidification process of alloys in a given system may be traced, using data both for the liquidus of the system and for the solid state regions. The data may be conveniently presented as a projected view of the system on the base of the space model. Fig. 3.4 shows the projected

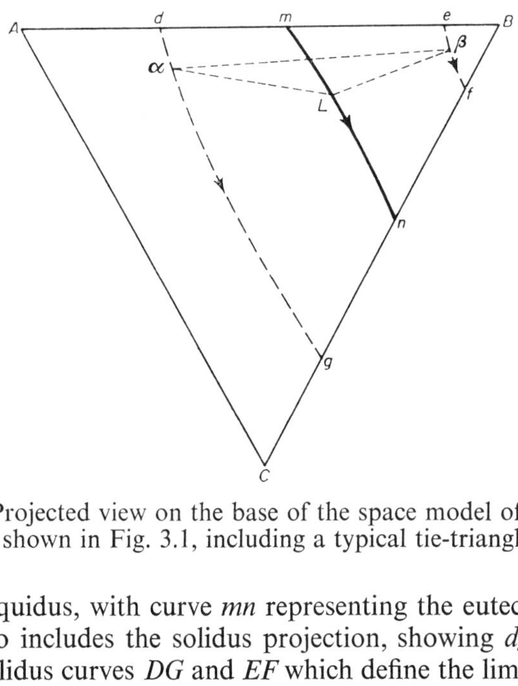

Fig. 3.4. Projected view on the base of the space model of the system shown in Fig. 3.1, including a typical tie-triangle

view of the liquidus, with curve *mn* representing the eutectic valley *MN*; the figure also includes the solidus projection, showing *dg* and *ef* representing the solidus curves *DG* and *EF* which define the limits of the α and β solid solution regions and of the $\alpha + \beta$ region at the completion of solidification. The α and β liquidus surfaces are represented by the regions *AmnC* and *Bmn* respectively, and the corresponding α, β, and $\alpha + \beta$ solidus surfaces by *AdgC*, *Bef*, and *defg* respectively.

If there are no solubility changes after the completion of solidification, the positions of the boundaries of the solid state regions in the projected view (Fig. 3.4) are the same as those shown in an isothermal section through the solid state portion of the system (Fig. 3.7*b*). In practice, changes of solid solubility, after solidification is completed, will generally cause the positions of the boundaries to differ in the two types of diagram, but it is sometimes convenient for purposes of discussion to ignore these changes.

The position of a typical tie-triangle is also shown in Fig. 3.4, but it should be appreciated that whereas a tie-triangle is isothermal in nature, the curves *mn*, *dg*, and *ef* are projections of curves extending over a range of temperature, as indicated by the arrows.

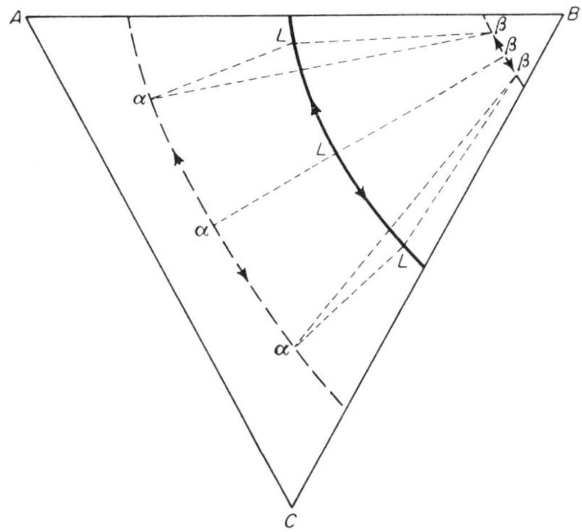

Fig. 3.5. Projected view of system in which the eutectic valley shows a maximum

The projected view of the system may be used as follows to trace the solidification processes of various alloys. The location of the alloy composition point either in region *AmnC* or in region *Bmn* defines whether α or β will be the primary phase during solidification; if the alloy composition lies on curve *mn* then solidification occurs directly by the eutectic reaction without the separation of primary α or β. Without experimental data for a given system, it is not possible to deduce from the diagram the composition of the primary phase at the beginning of solidification, or the precise changes in composition of the liquid and solid during the primary stage of solidification. However, if the primary β solid solubility is very limited the resultant changes in the liquid composition may be estimated by the procedure discussed later (see page 47).

If the alloy composition lies within one of the single-phase regions *AdgC* or *Bef*, then solidification is completed in the manner described for a system showing complete solubility. For alloys lying in the two-phase region *defg*, the liquid composition reaches curve *mn* during solidification and with further cooling, the eutectic reaction proceeds until the liquid is consumed.

The possibility of the eutectic curve showing a maximum or minimum has already been mentioned. In a projected view these cases are distinguished by the directions of the arrows on the projection of the eutectic curve (e.g. as in Figs. 3.5 and 3.6).

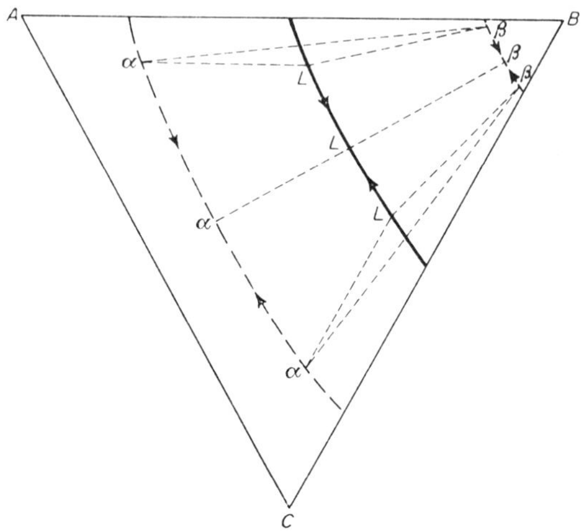

Fig. 3.6. Projected view of system in which the eutectic valley shows a minimum

At the maximum or minimum points, the tie-triangles reduce to a tie-line joining the compositions of L, α, and β in equilibrium. Detailed diagrams are given by Rhines.[7]

Isothermal sections

A representative isothermal section through the space model shown in Fig. 3.1 at a temperature between that of the eutectics in systems AB and BC is shown in Fig. 3.7*a*. The triangular section (i.e. tie-triangle) through the $L + \alpha + \beta$ region is seen in relation to the surrounding single-phase and two-phase fields. Typical tie-lines are shown traversing the two-phase regions.

Fig. 3.7*b* shows an isothermal section in the solid state at a temperature just below that of eutectic *BC*, and, as previously mentioned, if solid solubility changes after the completion of solidification are ignored, the positions of the boundary curves are identical with those in the solidus projection of Fig. 3.4.

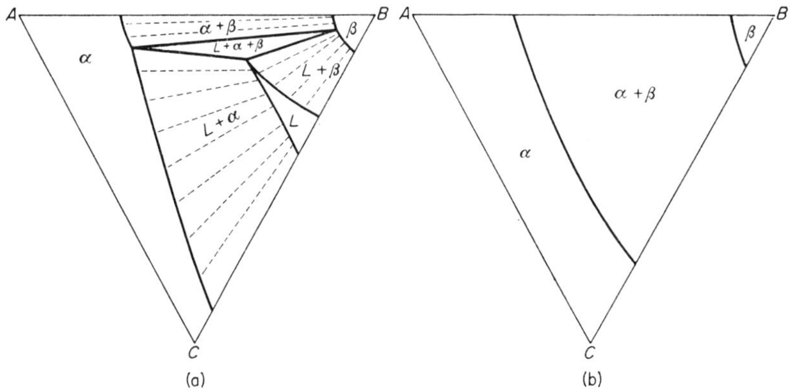

Fig. 3.7. Representative isothermal sections through the space model of Fig. 3.1

(*a*) at a temperature between that of the eutectics in systems *AB* and *BC*
(*b*) at a temperature below that of eutectic *BC*

Vertical section

The form of a vertical section passing through corner B and traversing the system to the midpoint *r* of side *AC* is shown in Fig. 3.8. The occurrence of the eutectic reaction over a range of temperatures is indicated by the three-phase region, $L + \alpha + \beta$. It should be noted that the boundaries of this region do not necessarily appear as straight lines, since the boundaries are sections through surfaces formed by series of tie-lines which do not generally lie in the plane of the section; further, in the two-phase regions, the tie-lines are not generally in the plane of the section.

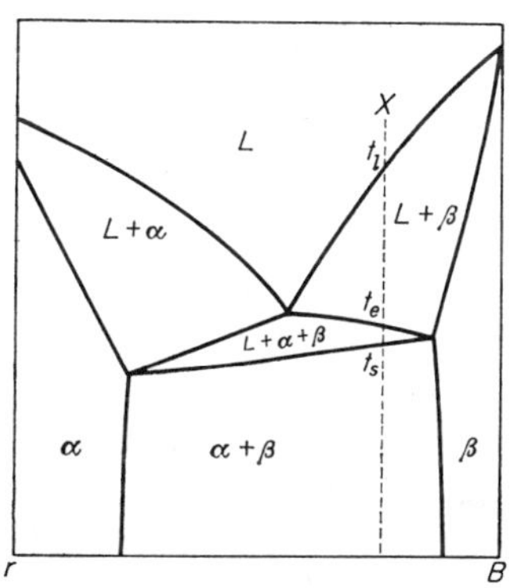

Fig. 3.8. Vertical section of system shown in Fig. 3.1 from corner *B* to the midpoint *r* of side *AC*

While this type of section does not provide information on the compositions of phases co-existing in equilibrium, it does show the temperatures at which phase changes occur in given alloys, and also the stages of solidification, e.g. for alloy X, t_l is the liquidus, t_e is the temperature at which the eutectic crystallization begins, and t_s is the temperature at which solidification is completed.

Systems containing a peritectic reaction $L + \alpha \longrightarrow \beta$

The hypothetical system ABC depicted in Fig. 3.9, represents complete liquid miscibility; the binary system BC shows complete solid solubility, while systems AB and AC involve partial solid solubility, and are of the peritectic type. The solid solutions are termed α and β respectively, of

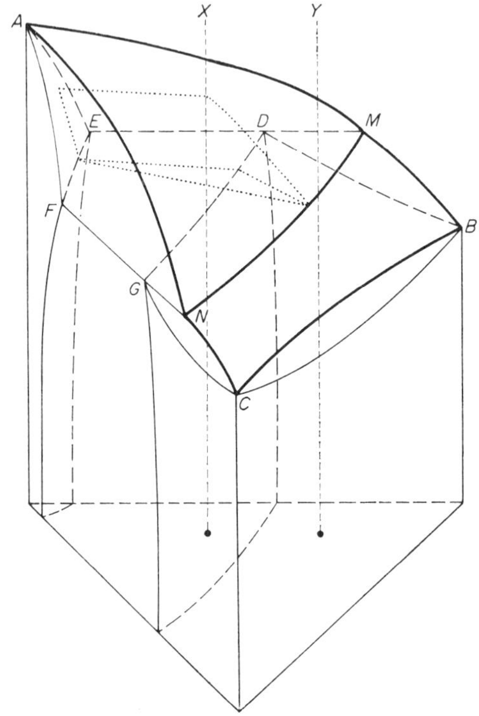

Fig. 3.9. Space model of system showing a peritectic reaction $L + \alpha \longrightarrow \beta$

which the latter extends completely across the BC side of the system. The temperature of the peritectic reaction $L + \alpha \longrightarrow \beta$ in system AB is greater than that of the same reaction in system AC.

The liquidus is seen to consist of two surfaces, *AMN* and *MBCN*, corresponding to the primary separation of α and β respectively, during solidification. These surfaces intersect along the peritectic curve *MN*, which is not a valley as in the eutectic type of system, but is rather like a 'shoulder'. The curves *EF* and *DG* join the points representing the respective compositions of α and β taking part in the peritectic reaction in the binary systems (these reactions are represented by the horizontal straight lines *EDM* and *FGN*). Surface *AEF* is the solidus of the liquid + α region, and *DBCG* is the solidus surface of the *L* + β region. The curves *MN*, *DG*, and *EF* do not lie in the same plane, and curve *DG* lies below the surface *EMNF*. The three-phase space, representing the co-existence of liquid, α and β taking part in the peritectic reaction, is bounded by the three surfaces *EMNF*, *DMNG*, and *EDGF*, the last mentioned being the solidus for the α + β part of the system; each of these surfaces is made up of tie-lines, namely *L*α, *L*β, and αβ tie-lines respectively. A vertical section through the three-phase space is shown in Fig. 3.16. Where the space terminates in the binary systems, it shrinks to become a line — either line *EDM* or *FGN*. (The nature of a system in which the three-phase space closes within the system is discussed later, page 36).

The peritectic reaction

Fig. 3.10 illustrates the form of the three-phase space and shows some of the representative tie-triangles that constitute the space. As the peritectic reaction proceeds with falling temperature, the liquid composition changes along curve *MN*, and the compositions of α and β change along *EF* and *DG* respectively.

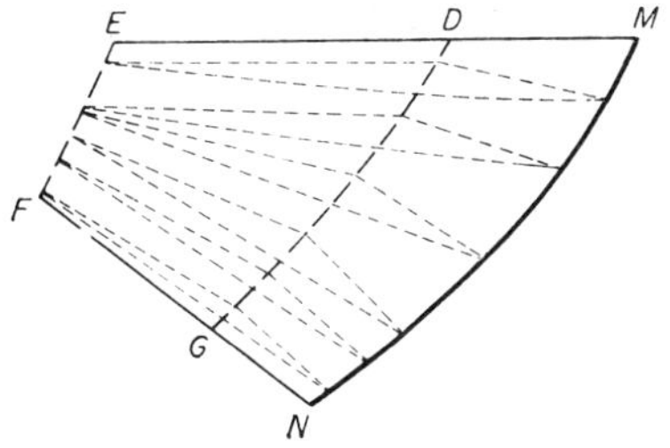

Fig. 3.10. Three-phase region (liquid, α and β) of Fig. 3.9

The process of peritectic solidification may be illustrated by reference to alloys lying in the region *EMNF*. These alloys commence solidification by the separation of crystals of α-phase, as, for example, in alloy X of Fig. 3.9; the liquid composition changes along a curved path on the liquidus surface, in the typical manner previously described, while the composition of α changes along the α-solidus surface. At some temperature, say t_1, the liquid composition reaches a point on the peritectic curve *MN* and simultaneously the α composition reaches a point on curve *EF*. At this temperature the *L*α tie-line forms one of the boundaries of the *L*αβ tie-triangle. Further cooling brings the alloy into the three-phase region *L* + α + β, and the peritectic reaction begins to occur. During the reaction either the liquid phase or the α phase may be consumed, depending on whether the alloy composition lies in the region *EDGF* or *DMNG* (corresponding to typical alloys such as 1 and 2 in the binary system shown in Fig. 3.11); these two cases in the ternary system may be illustrated by reference to the tie-triangles in a manner analogous to that discussed for the eutectic reaction in the previous section. Consider first alloy X lying in the region *EDGF*, that is, an alloy which consists of α and β phases when solid. As described above, primary solidification of α occurs until temperature t_1 is reached, when the situation is as illustrated in Fig. 3.12*a*; the alloy composition point lies on the $L\alpha$ tie-line forming one side of the $L\alpha\beta$ tie-triangle, and a slight decrease in temperature to t_2 will lead to the formation of some β by the reaction $L + \alpha \longrightarrow \beta$. The tie-triangle at t_2 is shown in Fig. 3.12*b* and the alloy composition point *X* is now within the triangle. At a lower temperature t_3 (Fig. 3.12*c*), the composition

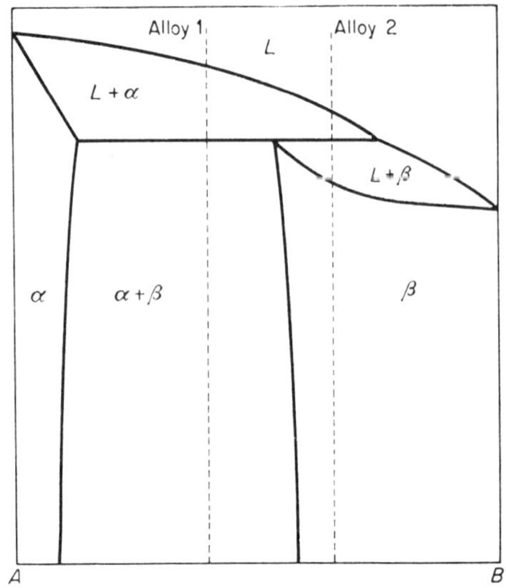

Fig. 3.11. Binary peritectic system

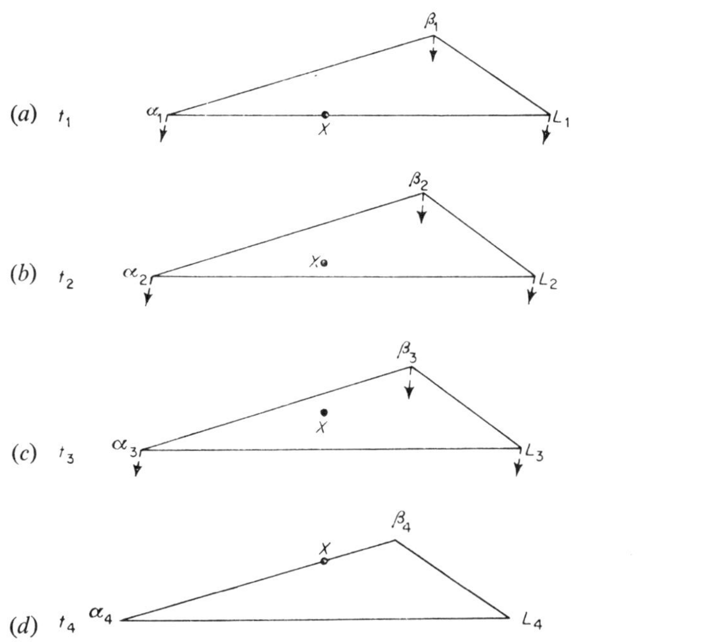

Fig. 3.12. Tie-triangles at various temperatures illustrating the progress of the peritectic reaction in an alloy, X, that is α + β when solid

point lies nearer to the αβ side of the triangle. The peritectic nature of the reaction, viz., $L + \alpha \longrightarrow \beta$ may be demonstrated by applying the 'centre of gravity' principle to the tie-triangles to calculate the percentages of the phases present at t_2 and t_3. It is found that the amount of liquid and the amount of α have decreased during cooling from t_2 to t_3, while the amount of β has increased. At t_4, the solidus temperature for the alloy, the alloy composition point lies on the αβ side of the tie-triangle (Fig. 3.12*d*), indicating that all the liquid has been consumed in the peritectic reaction.

For an alloy such as Y, lying in the region *DMNG*, primary solidification of α is followed by the peritectic reaction, but in this case the α phase is consumed during the peritectic reaction and *L* and β remain. The situation at the temperature at which the peritectic reaction is completed is depicted by the tie-triangle in Fig. 3.13, the alloy composition point *Y* now lying on the *L*β side of the triangle. With further cooling solidification continues by the separation of β, and the change in liquid composition occurs on a curved path along the *MBCN* portion of the liquidus; (the liquid composition is no longer restricted to movement along curve *MN*, since only one solid phase now exists in equilibrium with the liquid). Solidification is completed in the manner characteristic of solid solution alloys, and the final structure of the alloy is β solid solution.

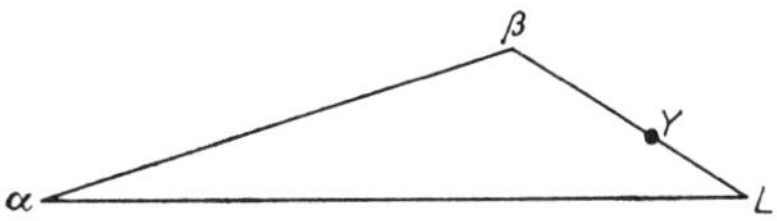

Fig. 3.13. Tie-triangle illustrating the completion of the peritectic reaction in an alloy Y, that consists only of β-phase when solidification is finally completed

Projected view of the system

A projected view of the ternary system is shown in Fig. 3.14. Curve *mn* is the projection of the liquidus curve *MN*, *Amn* is the region of primary α solidification, and *mBCn* is the region of primary β solidification. *dg* and *ef* are the projections of the solidus curves *DG* and *EF*, and the α and β solidus surfaces are indicated by the regions *Aef* and *dBCg*, which are the α and β solid solution regions at the completion of solidification. The α + β region is *edgf*. The position of a typical tie-triangle is indicated in the diagram.

The projected view of the system may be used to trace the solidification of alloys in a manner analogous to that described for the eutectic system in the previous section. The regions representing the projection of the liquidus are used to determine which phase will separate first during solidification, and the regions defined by the projections of the solidus curves *dg* and *ef* are used to determine the final solid state constitution. Alloys lying in the region *emnf* undergo the peritectic reaction, since the liquid composition reaches a point on curve *mn* as a result of α separation; if

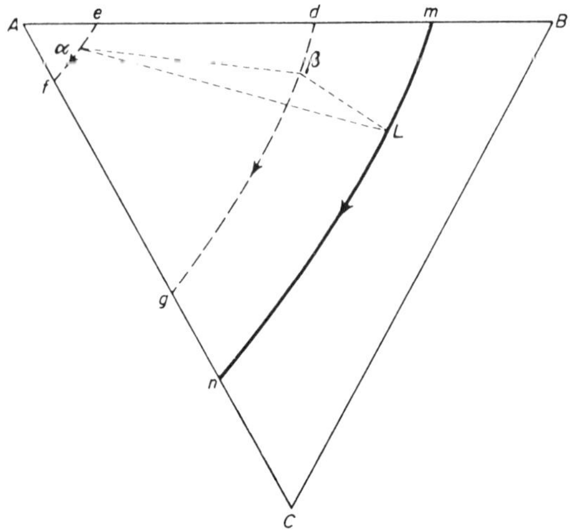

Fig. 3.14. Projected view on the base of the space model of the system shown in Fig. 3.9, including a typical tie-triangle

the alloy is in region *edgf*, solidification is completed at the end of the peritectic, the liquid being consumed and α + β remaining; within region *dmng*, the α is consumed, and solidification is completed by the direct separation of β from liquid giving a final structure of β solid solution.

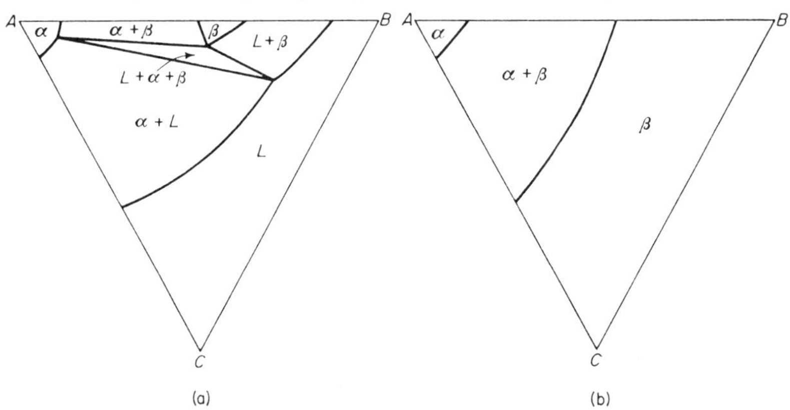

Fig. 3.15. Representative isothermal sections through the space model of Fig. 3.9
(*a*) at a temperature between that of the peritectic reactions in systems *AB* and *AC*
(*b*) at a temperature below that of the peritectic reaction in system *AC*

Isothermal sections

Fig. 3.15*a* shows a representative isothermal section through the space model at a temperature between that of the peritectic reactions in systems *AB* and *AC*; the $L + \alpha + \beta$ region is seen in relation to the adjoining regions.

Fig. 3.15*b* shows an isothermal section at a temperature below that of the peritectic reaction in system *AC*; if solid solubility changes after the completion of solidification are ignored, then the boundaries in the isothermal section correspond to those in the solidus projection (Fig. 3.14).

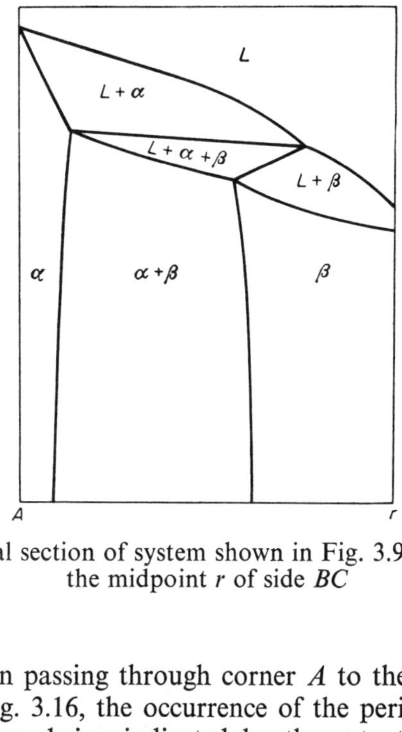

Fig. 3.16. Vertical section of system shown in Fig. 3.9 from corner *A* to the midpoint *r* of side *BC*

Vertical section

A vertical section passing through corner *A* to the midpoint *r* of side *BC* is shown in Fig. 3.16, the occurrence of the peritectic region over a range of temperature being indicated by the extent of the $L + \alpha + \beta$ region. The comments made previously on vertical sections apply here also.

Transition between eutectic and peritectic three-phase reactions

In the preceding sections the occurrence of eutectic and peritectic reactions has been discussed on the basis of tie-triangles that define the compositions of the reacting phases at various temperatures. A criterion for distinguishing between the two types of reaction is commonly based on a consideration of the tie-triangles in relation to their displacement, with falling temperature, along the curve representing the change of liquid composition. This criterion will be briefly presented here, although it should be noted that it is not satisfactory under certain conditions of compositional changes during the reaction; the reasons for the limitations of the criterion are not described here, but have been discussed by Hillert.[10]

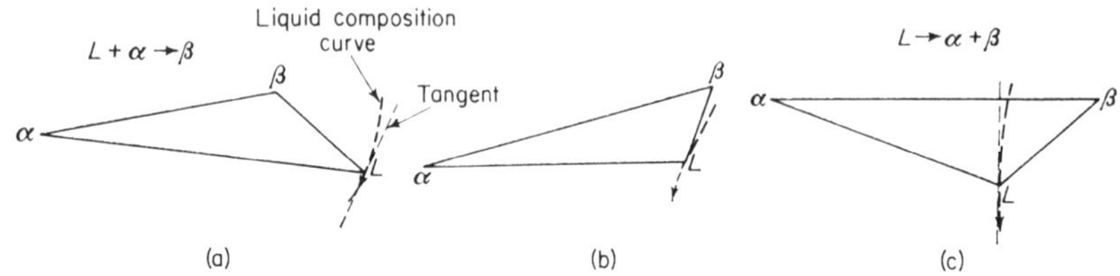

Fig. 3.17. Projected views of liquid composition curve (shown dashed) in relation to a tie-triangle for the peritectic and eutectic cases and for the transition case

In Figs. 3.17*a* and *c* the projected view of the liquid composition curve is shown in relation to the tie-triangle for the peritectic and eutectic cases respectively. In Fig. 3.17*a* at the temperature considered, the tangent drawn to the liquid composition curve, at the liquid composition corner of the tie-triangle, lies outside the triangle, and therefore does not cut the αβ side of the triangle; the reaction that is proceeding is taken to be peritectic in nature and with falling temperature displacement of the triangle is associated with the reaction $L + \alpha \longrightarrow \beta$. Fig. 3.17*c* illustrates the

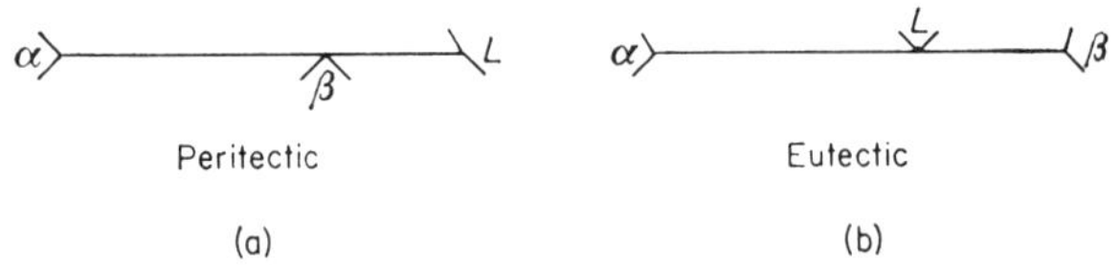

Fig. 3.18. Schematic representation of a peritectic and eutectic reaction in binary systems

eutectic case, in which the tangent to the liquid composition curve cuts the αβ side of the tie-triangle; then with decreasing temperature the liquid deposits a mixture of two solids i.e. $L \longrightarrow \alpha + \beta$. Figure 3.17*b* depicts a situation in which, as a transition stage, the $L\beta$ tie-line is tangential to the liquid composition curve at the liquid composition corner of the triangle.

This differentiation between the peritectic and eutectic examples may be compared with that made on the basis of the relative compositions of the phases involved in these types of reactions in binary systems. (Fig. 3.18.)

It is possible that within a given ternary system a transition may occur from one type of three-phase reaction to another, e.g. from peritectic to eutectic. Such an occurrence is illustrated by the ternary system shown in Figs. 3.19 and 3.20, in which the direction of the liquid composition curve for the three-phase reaction changes relative to those of the α and β composition curves within the system in such a way as to lead to a transition. Binary system BC is assumed to show complete solid solubility, AB is a peritectic system $L + \alpha \longrightarrow \beta$, and AC is a eutectic system $L \longrightarrow \alpha + \beta$.

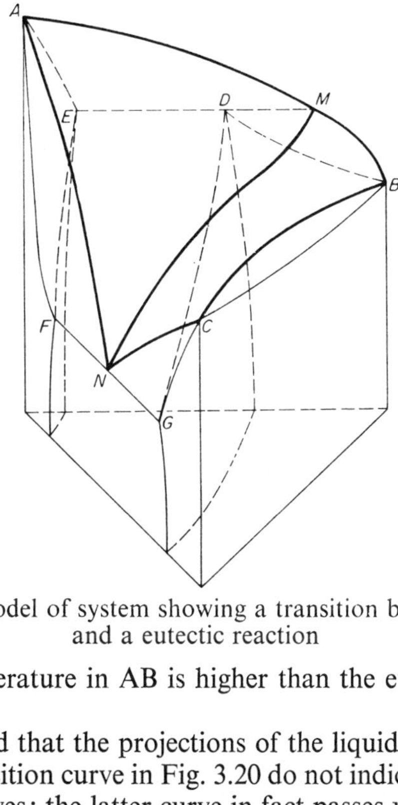

Fig. 3.19. Space model of system showing a transition between a peritectic and a eutectic reaction

The peritectic temperature in AB is higher than the eutectic temperature in AC.

It should be noted that the projections of the liquid composition curve and of the β composition curve in Fig. 3.20 do not indicate an actual intersection of these curves; the latter curve in fact passes under the former in the space model.

The three-phase region in the system is composed of a continuous series of tie-triangles of various shapes, part of the region being associated with the peritectic reaction and part with the eutectic reaction. Of the alloys that undergo the three-phase reaction, some, with compositions lying towards side *AB* of the system, will be clearly wholly peritectic in their three-phase stage of solidification, and others lying towards *AC* will be wholly eutectic; there is, however, an intermediate range of compositions in which there occurs a change in the nature of the three-phase reaction during the course of solidification.[10]

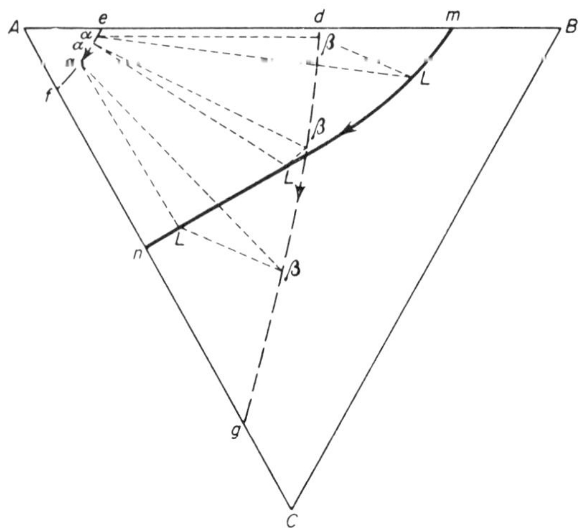

Fig. 3.20. Projected view of system of Fig. 3.19

If the temperature of the eutectic in system AC was higher than that of the peritectic in AB then a transition from eutectic to peritectic could occur within the ternary system.

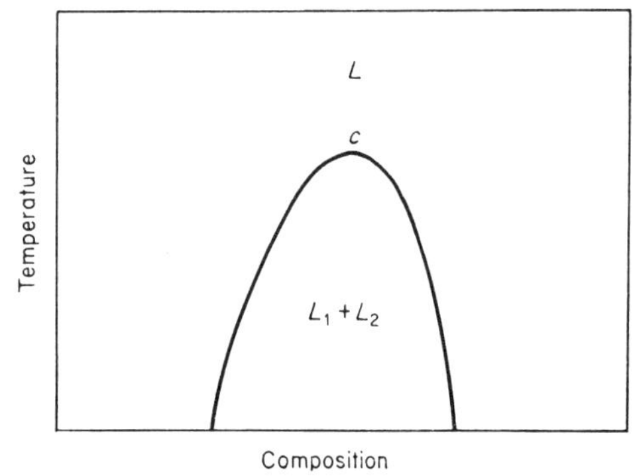

Fig. 3.21. Liquid miscibility gap in a binary system showing critical point, *c*

Systems containing a monotectic reaction $L_1 \longrightarrow L_2 + \alpha$

The presence of a monotectic reaction in a system is associated with a miscibility (or solubility) gap in the liquid state. The concept of a miscibility gap may be illustrated by reference to the portion of binary diagram shown in Fig. 3.21; there is partial solubility in the liquid state, and the

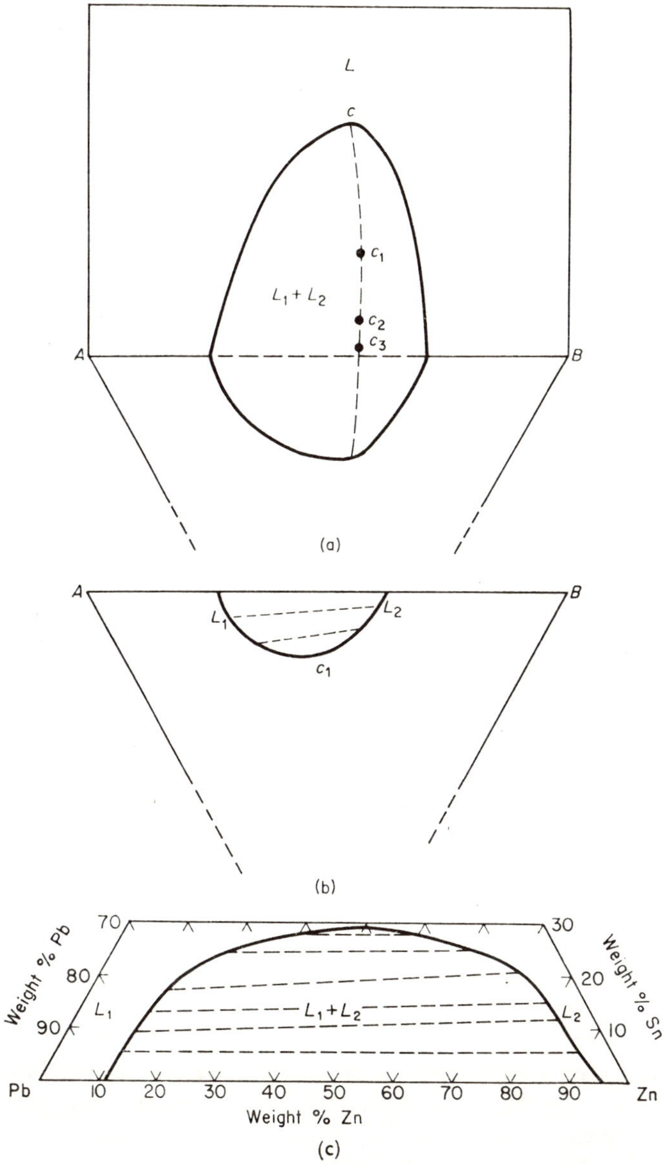

Fig. 3.22 (*a*). Liquid miscibility in a ternary system showing critical curve $cc_1c_2c_3$

(*b*) Isothermal section through the miscibility gap of Fig. 3.22*a* showing tie-lines

(*c*) Part of the isothermal section at 520°C of the Pb–Zn–Sn system showing tie-lines in the liquid miscibility gap. (From *Metals Handbook*, 7th Edn., 1948).

region in which two liquids co-exist is termed the miscibility gap. In the case shown, as the temperature increases the tie-lines spanning the miscibility gap become shorter, until, at the temperature of the critical point, *c*, the tie-line has zero length and the gap closes; above this temperature there is complete liquid solubility.

An analogous case for a ternary system is shown in Fig. 3.22*a* where a liquid miscibility gap in binary system AB extends into the ternary system to form a region that has the shape of a partial dome. It will be seen that the miscibility gap closes within the ternary system, and that outside the gap there is complete liquid solubility.

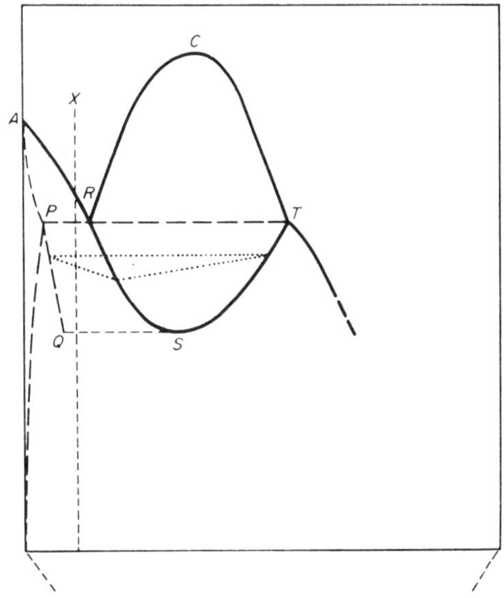

Fig. 3.23. View of part of the space model of a ternary system containing a miscibility gap giving rise to a monotectic reaction

An isothermal section through the miscibility gap is shown in Fig. 3.22*b*; the diagram indicates the positions of typical L_1L_2 tie-lines and of the critical point c_1 at which the tie-line reaches zero length. (Fig. 3.22*c* shows the miscibility gap in an isothermal section of the Pb–Zn–Sn system.[9]) For each temperature considered there is a similar critical point, and these points constitute the critical curve $cc_1c_2c_3$, etc., shown in Fig. 3.22*a*. An analogous type of critical curve exists for the case of a solid state miscibility gap that closes within a ternary system (see page 36).

The example shown is not the only possible form for a liquid miscibility gap; for example, there may be a critical point within the system at a temperature higher than the binary critical point.

A liquid miscibility gap can give rise to a monotectic type of reaction, and Fig. 3.23 shows a part of a ternary system in which one of the binary systems, AB, shows the reaction

$$\text{Liquid}_1\ (L_1) \longrightarrow \text{Liquid}_2\ (L_2) + \alpha \text{ solid solution}$$

The partial dome representing the $L_1 + L_2$ region lies upon the liquidus surface of the ternary system; the critical point *S* lies at a lower temperature than that of the binary monotectic reaction. There exists a three-phase space representing the co-existence of L_1, L_2, and α solid solution, the space being made up of $L_1L_2\alpha$ tie-triangles, and the occurrence of the reaction $L_1 \longrightarrow L_2 + \alpha$ can be considered in terms of these tie-triangles. As the

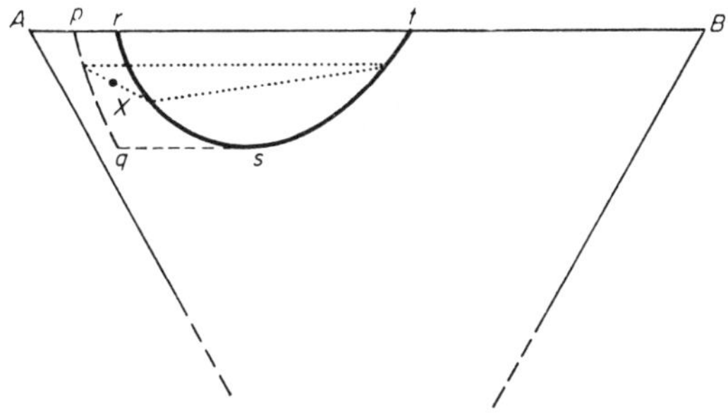

Fig. 3.24. Projected view of Fig. 3.23, including a tie-triangle

reaction occurs with falling temperature, the composition of L_1 changes along *RS*, that of L_2 along *TS*, and that of α along *PQ*. Within the ternary system the three-phase space terminates at the tie-line *QS*.

The solidification of a typical alloy, X, may be used to illustrate the monotectic reaction (Figs. 3.23 and 3.24). Solidification begins by the separation of primary α, and the compositional changes that occur with falling temperature eventually bring the liquid composition on to curve *RS* and the α-composition on to curve *PQ*, at some temperature t_1; the equilibrium at t_1 is illustrated by the tie-triangle in Fig. 3.25*a* in which the alloy composition point lies on the $L_1\alpha$ side of the triangle. At a lower temperature, t_2 (Fig. 3.25*b*), the alloy composition point *X* now lies within the tie-triangle, and three phases exist, namely α, L_1, and L_2. At a still lower temperature, t_3 (Fig. 3.25*c*), *X* lies on the αL_2 side of the triangle, indicating that none of L_1 remains. Thus, at t_3, the monotectic reaction is completed for the alloy X, and on further cooling, L_2 (now no longer constrained to move along curve *TS*) changes composition along a curved path on the liquidus surface, depositing α solid solution. The further course of solidification depends on the nature of the remaining portion of the system, not shown in Fig. 3.23.

For a fuller description of the monotectic reaction, reference may be made to other accounts[4,5] where it is assumed that solid solubility is negligible, and for this condition, it is shown how either of the two liquids (i.e. either L_1 or L_2) may be consumed during the reaction, depending on the alloy composition.

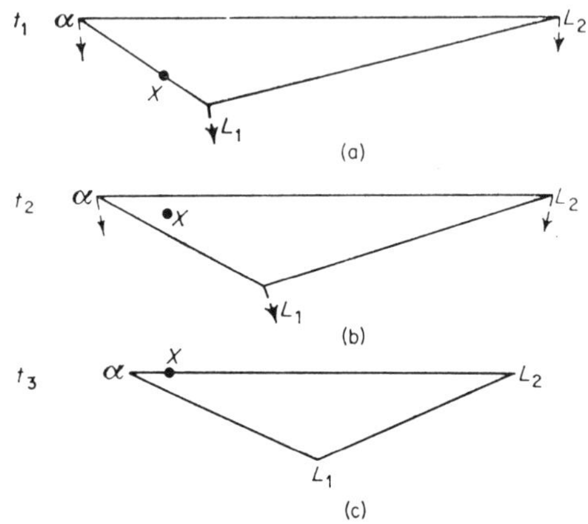

Fig. 3.25. Tie-triangles illustrating the progress of the monotectic reaction in an alloy X

Systems containing a solid state miscibility gap that closes within the system

In the eutectic and peritectic systems previously discussed, the $\alpha + \beta$ region may be referred to as a solid state miscibility gap that extends across the system. In some systems, a two-phase region of $\alpha + \beta$ may not span the system completely, but may close within the system in a manner essentially analogous with the case of the liquid miscibility gap. Two examples are illustrated in Figs. 3.26–3.29; the first example is a ternary system in which binary system AB is eutectic in nature, showing partial solid solubility, while binary systems BC and AC show complete solid solubility; in the other example, system AB is assumed to be peritectic in nature. Both of these ternary systems contain a solid state $\alpha + \beta$ region which extends only part of the way across the system, and isothermal sections through the region at, and below the temperature of point N will show a critical point. The three-phase space ($L + \alpha + \beta$) which rests on the top of the $\alpha + \beta$ region, also closes within the system.

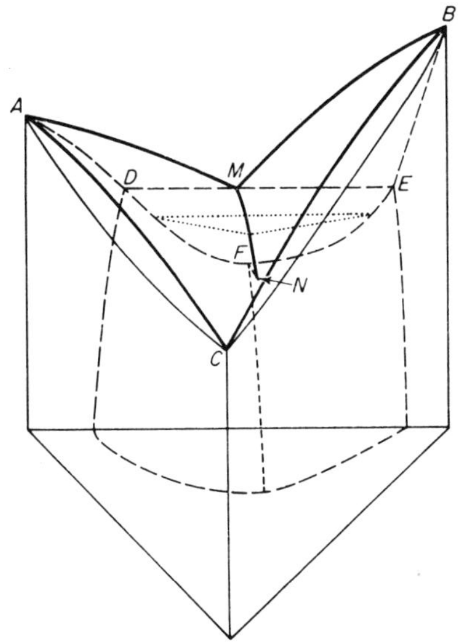

Fig. 3.26. Space model of a system containing a eutectic reaction in which the solid state miscibility gap closes within the system

The liquidus curves giving the compositions of liquid involved in either the eutectic or the peritectic reactions are shown in relation to the tie-triangles, representing the co-existence of L, α, and β. The changes of α

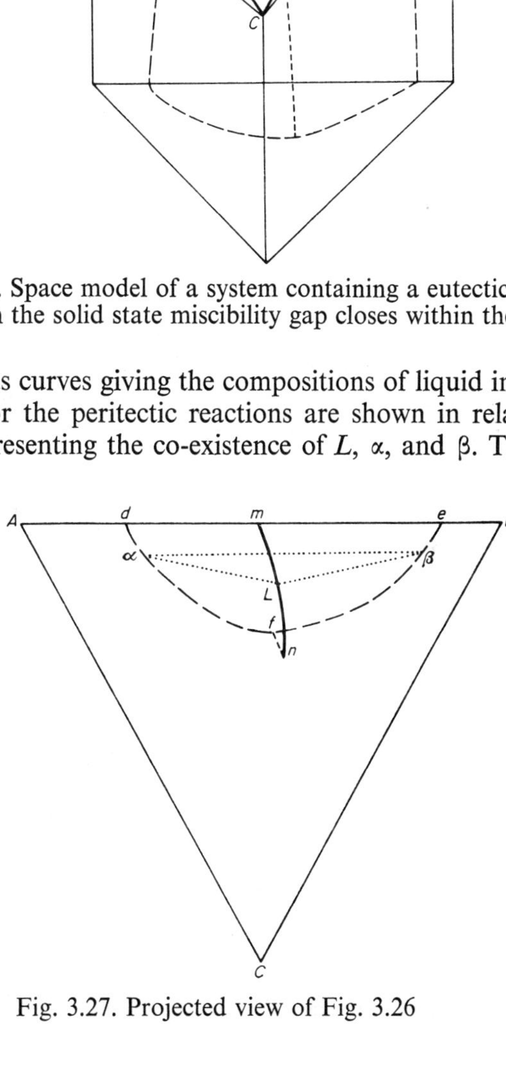

Fig. 3.27. Projected view of Fig. 3.26

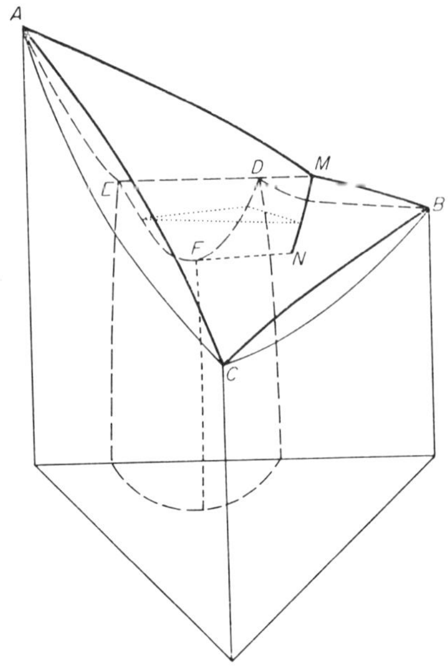

Fig. 3.28. Space model of a system containing a peritectic reaction in which the solid state miscibility gap closes within the system

and β compositions as the three-phase reaction proceeds, occur in a manner analogous to the changes of the liquid compositions in a monotectic reaction, and the positions of tie-triangles relative to alloy composition

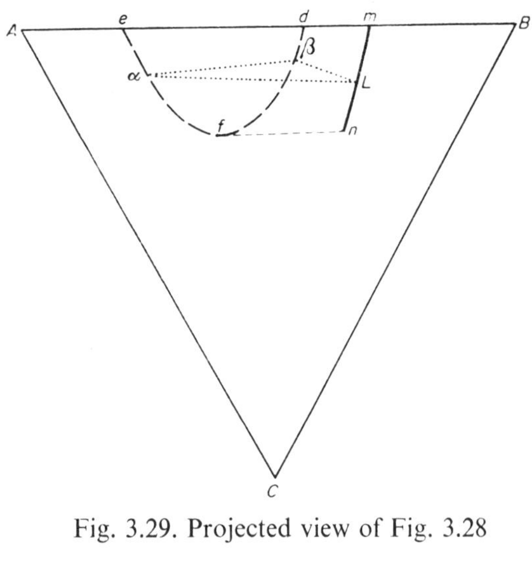

Fig. 3.29. Projected view of Fig. 3.28

points form a basis for discussing the progress and completion of the reaction. For example, in Fig. 3.26 during the eutectic reaction the liquid composition changes along curve *MN* while the compositions of the solid solutions change along *DF* and *EF* respectively. There are some special features of each of these systems regarding the shape of the three-phase space, and the solidification of certain alloys which, although single phase when solid, undergo the three-phase reaction during solidification. Reference should be made to other accounts for details.[3,4,7]

Problems

1. With reference to Fig. 3.1:

(a) State what phase regions are separated by the following surfaces: (i) *ADGC*, (ii) *BMN*, (iii) *DMNG*, (iv) *DEFG*.

(b) State what surfaces separate the following phase regions: (i) $L + \beta/\beta$, (ii) $L + \beta/\alpha + \beta + L$.

2. In the system ABC, a ternary alloy containing 40% B and 8% C undergoes a eutectic reaction $L \longrightarrow \alpha + \beta$ over the temperature range 550–500°C as part of its solidification sequence. The compositions of the phases coexisting in equilibrium at 540°C and 510°C are as follow:

	α	β	L
540°C	85%A, 10%B, 5%C	5%A, 93%B, 2%C	55%A, 30%B, 15%C
510°C	82%A, 11%B, 7%C	6%A, 89%B, 5%C	48%A, 32%B, 20%C

Calculate: (a) The proportions by weight of α and β phases present in the alloy at 540°C.

(b) The ratio of the proportions of liquid phase present at 540°C and 510°C.

3. (a) With reference to Fig. 3.8 describe the solidification sequence of alloy X; refer also to Fig. 3.1 to describe the compositional changes that occur during solidification.

(b) On what surfaces respectively of Fig. 3.1 do the points t_e and t_s shown in Fig. 3.8 lie?

4. A ternary system ABC shows complete solubility in the liquid state and contains only two solid phases, which are solid solutions designated α and β respectively. The melting points of components A, B and C are 1100°C, 900°C and 800°C, respectively. The binary system BC shows complete solid solubility, while systems AB and AC each contain an invariant reaction as follows:

	Temperature of invariant reaction	Compositions of the phases involved in invariant reaction		
		α	β	Liquid
System AB	850°C	95%A, 5%B	10%A, 90%B	50%A, 50%B
System AC	750°C	97%A, 3%C	12%A, 88%C	60%A, 40%C

Sketch and label:

(a) Possible liquidus and solidus projections for the ternary system.

(b) By reference to these projections, describe the solidification sequence of an alloy containing 30% A, 55% B and 15% C, under equilibrium conditions; estimate the proportions of the phases present at the completion of solidification.

5. With reference to Fig. 3.9:

(a) State what phase regions are enclosed by the following surfaces: (i) *AME*, *AEF*, *AMN*, *ANF* and *EMNF*; (ii) *MBD*, *MBCN*, *NCG* and *DBCG*; (iii) *EDGF*, *DMNG* and *EMNF*.

(b) State what surfaces separate the following phase regions: (i) $\alpha + \beta / L + \alpha + \beta$, (ii) $L / L + \alpha$.

6. A ternary system ABC shows complete liquid solubility and contains only two solid phases, namely two solid solutions designated α and β respectively.

The following data refer to the compositions of liquid, α and β coexisting in equilibrium at the temperatures shown:

Temperature (°C)	Compositions		
	Liquid	α	β
550	69% A, 19% B, 12% C	57% A, 41% B, 2% C	20% A, 78% B, 2% C
520	66% A, 18% B, 16% C	56% A, 40% B, 4% C	19% A, 78% B, 3% C
500	63% A, 17% B, 20% C	55% A, 39% B, 6% C	18% A, 78% B, 4% C

(a) Calculate the equilibrium percentages of liquid, α and β respectively, present in an alloy containing 44% A, 50% B and 6% C at 520°C.

(b) Deduce, giving your reasoning, the nature of the three phase reaction undergone by the alloy containing 44% A, 50% B and 6% C during solidification.

(c) Sketch a projected view of the liquidus, and an isothermal section in the solid state for a system consistent with the data given above: also sketch an isothermal section for 550°C.

(d) With reference to the sketches of (c), describe fully the solidification sequence you would expect for an alloy containing 60% A, 30% B and 10% C, assuming equilibrium conditions.

7. Making use of data obtained from Fig. 3.22*c*, calculate how the proportion of Pb-rich liquid will vary with Sn content in the range 0–25 wt% Sn in a series of Pb–Zn–Sn alloys each containing 55 wt% Pb, equilibrated at 520°C.

4

Systems Containing Four Phases

AN invariant reaction in a ternary system occurs when four phases co-exist in equilibrium. Two forms of such reaction will be discussed, each involving a liquid (L) and three solid phases (primary solid solutions) α, β, and γ:

(1) A ternary eutectic reaction

$$L \longrightarrow \alpha + \beta + \gamma$$

(2) A ternary peritectic reaction

$$\text{type } (a)\ L + \alpha \longrightarrow \beta + \gamma$$
$$\text{type } (b)\ L + \alpha + \beta \longrightarrow \gamma$$

In some accounts the designation 'peritectic' is restricted to the type (*b*) reaction, and type (*a*) is viewed as being intermediate in nature between eutectic and peritectic. However, in the present account, the term 'peritectic' is applied to both these types.

Each of the reactions listed is associated with the intersection of three liquidus curves, each liquidus curve representing the path along which the liquid composition changes during a three-phase reaction; the basic features of each of the invariant reactions are revealed by consideration of the liquidus curves and the associated tie-triangles.

Systems containing a ternary eutectic reaction

A system in which the three binary systems are of the eutectic type

Fig. 4.1 shows the space model of a ternary system ABC in which each of the constituent binary systems is of the eutectic type, with partial solid solubility of the components: the primary solid solutions are α, β, and γ respectively and no intermediate phases are present. Eutectic valleys (*DG*,

FG, and *EG*) run from the binary systems to lower temperatures to meet at the ternary eutectic point *G*. For each of these eutectic reactions there exists a three-phase region made up of tie-triangles. For the $L \longrightarrow \alpha + \beta$ reaction, curve *DG* represents the composition of the liquid, curve M_1M the composition of α, and N_1N the composition of β; for the $L \longrightarrow \beta + \gamma$

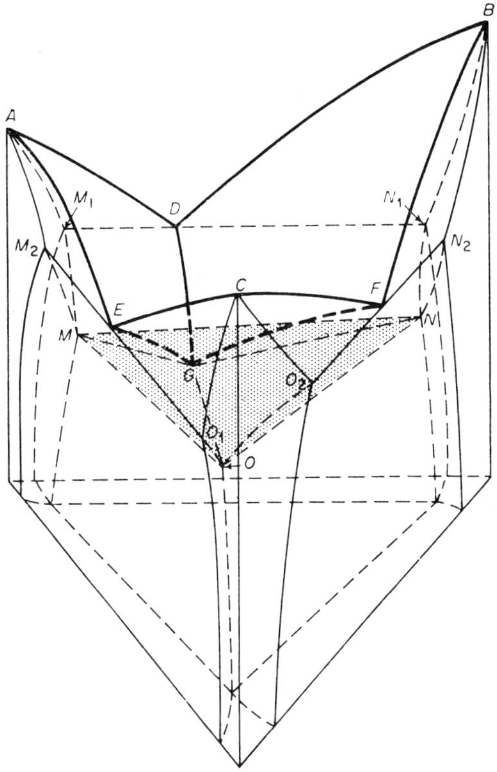

Fig. 4.1. Space model of system showing a ternary eutectic reaction $L \longrightarrow \alpha + \beta + \gamma$

reaction, the liquid composition lies on *FG*, and the β and γ compositions on N_2N and O_2O respectively, while for the $L \longrightarrow \alpha + \gamma$ reaction, the liquid composition lies on *EG* and the α and γ compositions on M_2M and O_1O respectively.

With decreasing temperature the positions of the tie-triangles change as the compositional changes occur in the phases involved in the reaction. The ternary eutectic reaction occurs at the temperature at which the liquid

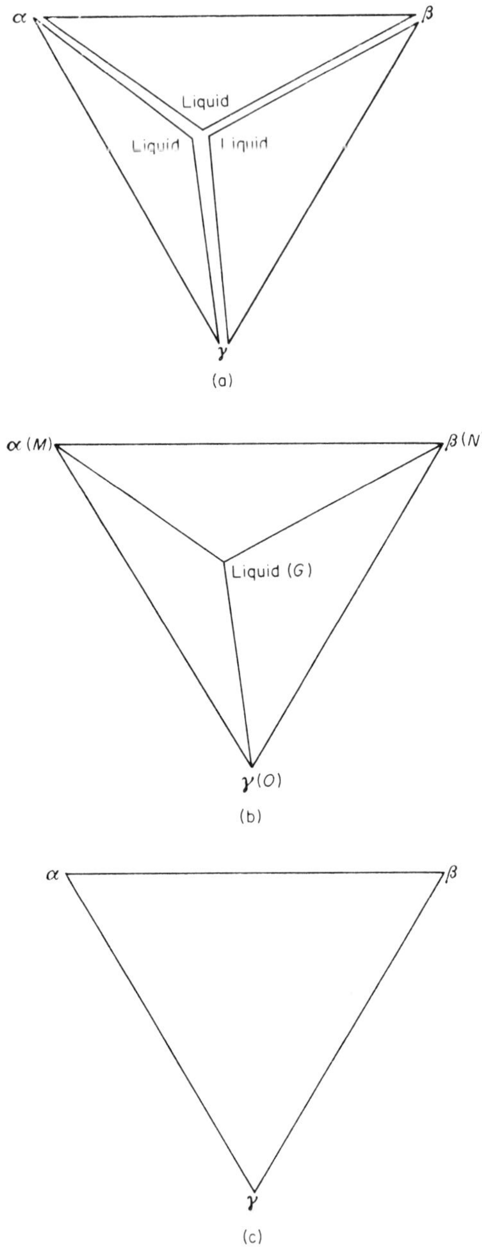

(*a*) at a temperature just above that of the ternary eutectic;

(*b*) at the ternary eutectic temperature;

(*c*) below the ternary eutectic temperature

Fig. 4.2. Tie-triangles

composition corners of these three triangles become coincident with one another at the ternary eutectic point (G in Fig. 4.1); at the invariant reaction temperature the $L\alpha$ sides of the triangles coincide, as do also the $L\beta$ and $L\gamma$ sides respectively. Fig. 4.2*a* indicates the relative positions of the tie-triangles at a temperature just above the ternary eutectic, and Fig. 4.2*b* shows the merging of the triangles at the ternary eutectic temperature. The ternary eutectic reaction is represented by a plane in the space model (*MNO* in Fig. 4.1); the composition of each of the four phases participating in the reaction is fixed, and is therefore represented by a point on the plane, and the liquid composition point lies within the triangle formed by the compositions of the three solid phases. Under equilibrium conditions, the temperature remains constant during the eutectic reaction until all the liquid is consumed.

Below the ternary eutectic plane there lies a three-phase region, $\alpha + \beta + \gamma$, and the equilibrium state at a temperature within this region is represented by the $\alpha\beta\gamma$ tie-triangle shown in Fig. 4.2*c*. During cooling in the solid state, solubility changes generally occur, and the composition co-ordinates of the α, β, and γ corners will therefore change; however, it is convenient to neglect these changes, although they are indicated in Fig. 4.1.

Thus, the ternary eutectic reaction involves three three-phase regions that approach the eutectic plane from higher temperatures, and one three-phase region existing below the plane.

Projected view of the system

Fig. 4.3 shows a projected view of the system. Curves *dg*, *fg*, and *eg* are the liquidus curves, the eutectic nature of the invariant reaction being indicated by the arrows on the curves all pointing towards *g*. Regions *Adge*, *Bdgf*, and *Cegf* define the primary phase solidification, i.e. of α, β, and γ respectively. The solidus curves m_1m and m_2m define the boundaries of the α-solid solution region at the completion of solidification, while the β and γ single-phase regions are bounded by n_1n and n_2n, and o_1o and o_2o respectively. The triangle *mno* is the projection of the isothermal eutectic plane and on its upper surface is divided into the three tie-triangles *mng*, *mgo*, and *ngo*; the tie-triangle *mno* defines the three-phase region in the solid state while the solid state constitution is completed by the two-phase regions of $\alpha + \beta$, $\beta + \gamma$, and $\alpha + \gamma$ respectively (m_1n_1nm, nn_2o_2o, and m_2moo_1).

If solid state solubility changes are neglected, Fig. 4.3 serves to define the limits of the solid state regions.

The solidification of alloys

The solidification of alloys in the various regions of the system may be discussed on the basis of Fig. 4.3 assuming equilibrium conditions. If an alloy lies in one of the single-phase regions, solidification occurs in a manner typical of a solid-solution type alloy as described in a previous section. For alloys lying in one of the two-phase regions but not on a eutectic valley, the primary phase solidification occurs, either α, β, or γ, depending on the primary field in which the alloy composition lies. As a

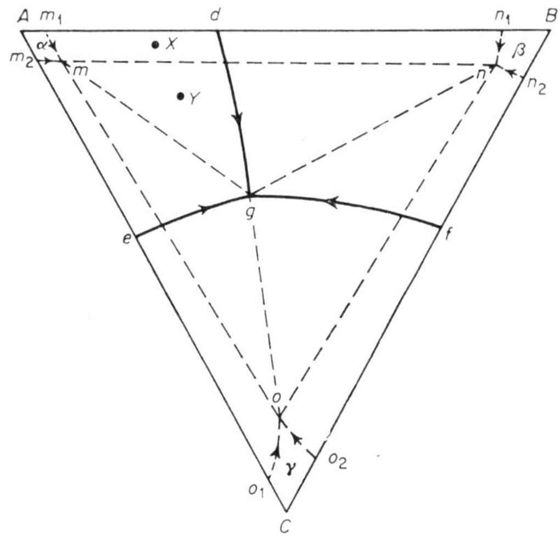

Fig. 4.3. Projected view of system of Fig. 4.1

result of the separation of the primary phase, the liquid composition reaches one of the eutectic valleys, and the appropriate reaction begins; solidification is completed with falling temperature, leaving a mixture of the two solid phases. In alloy X, for example, the primary separation of α is followed by the reaction $L \longrightarrow \alpha + \beta$: during the reaction the liquid corner of each tie-triangle lies on curve *dg*, and the α and β corners on curves m_1m and n_1n respectively, and solidification is complete when the αβ side of the tie-triangle contains point *X*.

If the alloy composition lies in the three-phase region, the liquid composition reaches the ternary eutectic point *g*, e.g. in alloy Y, as a result of the occurrence of the $L \longrightarrow \alpha + \beta$ eutectic; the $L\alpha\beta$ tie-triangle has moved down on to the ternary eutectic plane without the alloy composition point coming to lie on the αβ side of the triangle. At the ternary invariant temperature, the $L\alpha\beta$ tie-triangle is *mng*, and the reaction $L \longrightarrow \alpha + \beta + \gamma$

occurs, with the liquid composition represented by point *g*, and the α, β, and γ compositions by *m*, *n*, and *o* respectively.

In the type of system discussed, which shows partial solid solubility, the actual path of the liquid composition on the liquidus surface during the primary stage of solidification can only be determined experimentally. This aspect has been previously discussed.

However, if it is assumed that the system shows no solid solubility, the compositional changes which occur during solidification may be accurately traced. Fig. 4.4 represents the liquidus projection of such a system; all

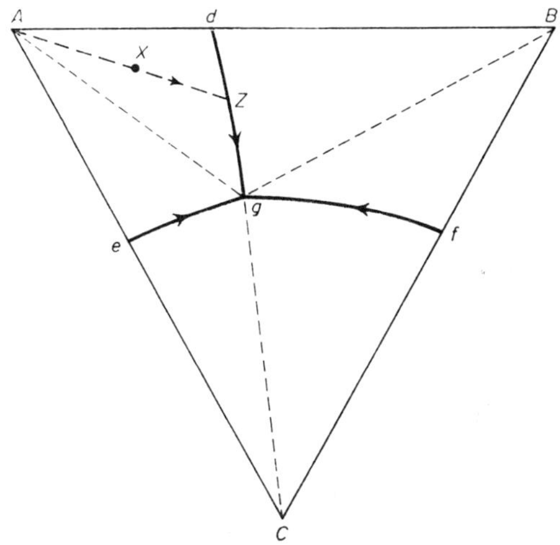

Fig. 4.4. Projected view of ternary eutectic system $L \longrightarrow A + B + C$

ternary alloys consist of A + B + C in the solid state. The regions of the primary solidification of A, B, and C respectively are *Adge*, *Bdgf*, and *Cegf* respectively.

The extremities of the tie-lines representing the liquid compositions at successively lower temperatures form the path of the liquid composition on the liquidus surface (Fig. 4.4). Since the composition of the primary solid at all temperatures is the same, the liquid composition path is, in fact, the extension of the straight line joining the composition of the solid and the alloy composition (e.g. *AX* in Fig. 4.4). When the liquid composition reaches a point on one of the binary eutectic valleys the primary solidification ceases (e.g. for alloy X, this occurs at point *Z* on curve *dg*). (At this temperature, prior to the commencement of the eutectic reaction, the amounts of liquid and solid co-existing can be calculated from the tie-line *AXZ*; thus, the amount of primary A is given by the ratio *XZ*/*AZ*.)

Then, with falling temperature the eutectic reaction commences and, as the two solids separate, the liquid composition moves down the eutectic valley (e.g. for alloy X, along portion *Zg* of curve *dg*, while A and B separate). The progress of the eutectic reaction can be represented by tie-triangles at decreasing temperatures (t_1, t_2, and t_3) in Fig. 4.5; since the compositions of the solids do not change during the reaction, the side of the triangle joining the two solids (i.e. *AB* in Fig. 4.5) cannot move so as

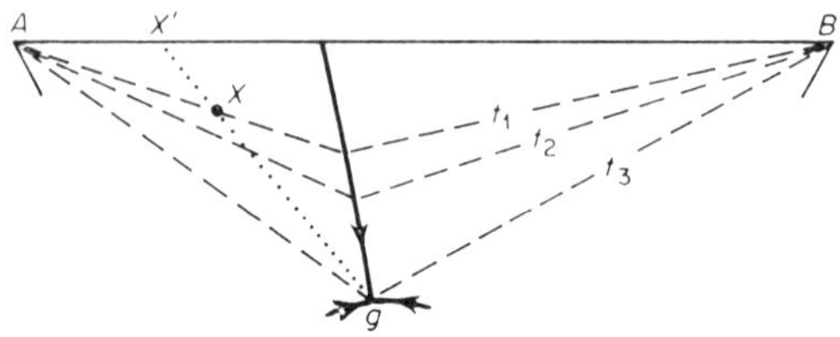

Fig. 4.5. Tie-triangles illustrating the progress of the eutectic reaction $L \longrightarrow A + B$ in the system of Fig. 4.4

to contain the alloy composition point *X*. Solidification cannot be completed until the liquid composition reaches the ternary eutectic point *g*, when the reaction $L \longrightarrow A + B + C$ occurs. (The percentage of liquid that finally solidifies as the ternary eutectic can be calculated from the tie-triangle *AgB* (Fig. 4.5): it is given by $\frac{XX'}{gX'} \times 100$.)

Referring to Fig. 4.4, if the alloy composition lies on one of the three straight lines joining the respective components to the ternary eutectic point (e.g. line *Ag*), the primary solidification stage is followed directly by the ternary eutectic reaction. Alloys in the region *Adg* undergo the eutectic reaction $L \longrightarrow A + B$ during solidification, while alloys in region *Aeg* undergo the reaction $L \longrightarrow A + C$: similarly, lines *Bg* and *Cg* divide the other primary fields in a corresponding way.

In systems where there is solid solubility, but very limited in extent, the liquid composition change may be estimated by the method described above for the case of no solid solubility, i.e. by drawing the extension of the straight line from the appropriate corner of the system through the alloy point. However, this procedure is only an approximation where solid solubility exists, and should not be used where the solid solubility is extensive.

Isothermal sections

Figs. 4.6*a* and 4.6*b* show two isothermal sections through the system with partial solid solubility (Fig. 4.1). The section shown in Fig. 4.6*a* is at a temperature just above the ternary eutectic temperature, but below the temperature of the eutectic reactions in the three binary systems. The three triangles representing the $L\alpha\beta$, $L\beta\gamma$, and $L\alpha\gamma$ regions are seen in

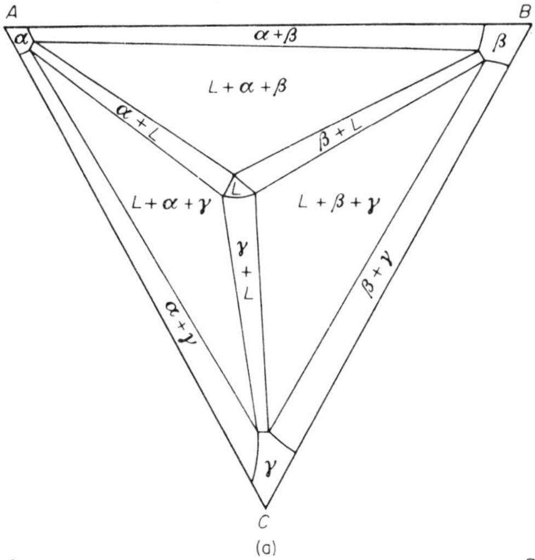

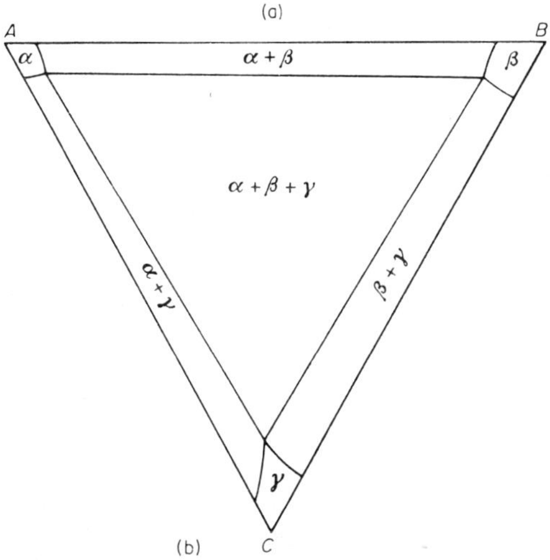

Fig. 4.6. (*a*) Isothermal sections through the system of Fig. 4.1 at a temperature above the ternary eutectic temperature, but below the temperatures of the eutectic reactions in the three binary systems
(*b*) at a temperature below that of the ternary eutectic

relation to the adjoining single-phase and two-phase regions; with decreasing temperature, the triangles move down on to the ternary eutectic plane as described earlier. Fig. 4.6*b* shows a section at a temperature just below that of the eutectic reaction; if solid state solubility changes are neglected, the limits of the solid phase regions in this isothermal section will coincide with those shown in the projected view (Fig. 4.3).

In the solid state isothermal section (Fig. 4.6*b*) thermodynamic considerations lead to the following principle regarding the boundaries between the single-phase and two-phase regions where they meet the three-phase region; if these boundaries are produced, then they must either both lie within the three-phase region, or must project into two different two-

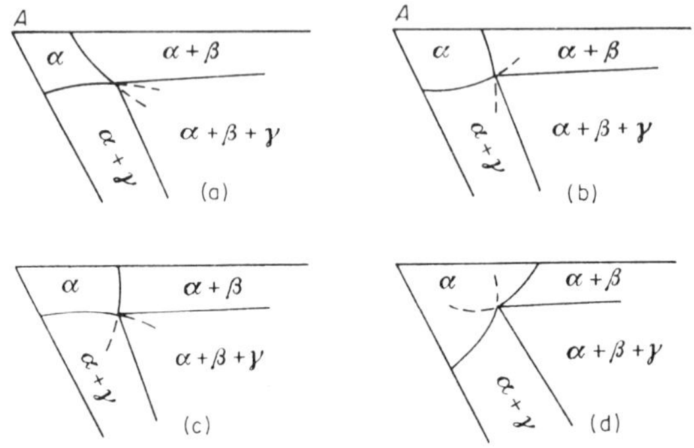

Fig. 4.7 (*a*) and (*b*) correct and (*c*) and (*d*) incorrect constructions for boundaries of single-phase regions

phase fields, as shown for example in Figs. 4.7*a* and *b*.[11] Incorrect constructions are shown in Figs. 4.7*c* and *d*; in the former case one of a pair of boundaries projects into the three-phase region and the other into a two-phase region, while in the latter both boundaries are shown as projecting into the single-phase region.

A system in which one of the binary systems is of the peritectic type and two are of the eutectic type

If one of the binary systems is of the peritectic type, the ternary invariant reaction can still be of the eutectic type; the binary peritectic may change to eutectic as it moves into the system. An example of this is discussed later with reference to a system containing an intermediate phase (see page 74).

Systems containing a ternary peritectic reaction

A system in which one binary system is of the peritectic type, and two are of the eutectic type, and which shows an invariant reaction of the type $L + \alpha \longrightarrow \beta + \gamma$

In the ternary system ABC illustrated in Fig. 4.8, the system AB is peritectic in nature, and systems BC and AC are eutectic in nature.

Terminal solid solutions α, β, and γ are formed, and there are no intermediate phases. (The type of invariant reaction depicted can also occur with other combinations of binary systems.[3])

The curve *DG* running into the ternary system represents the path of the composition of liquid taking part in the $L + \alpha \longrightarrow \beta$ peritectic reaction. Curve *EG* is the liquid composition curve for the eutectic reaction,

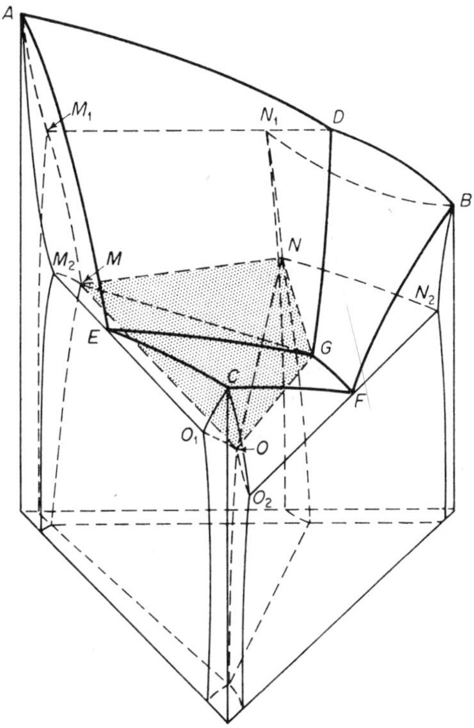

Fig. 4.8. Space model of system showing a ternary peritectic reaction
$L + \alpha \longrightarrow \beta + \gamma$

$L \longrightarrow \alpha + \gamma$. Point *G*, where these curves intersect, represents the composition of liquid taking part in the invariant peritectic reaction $L + \alpha \longrightarrow \beta + \gamma$. From *G*, the eutectic valley, *GF*, for the reaction $L \longrightarrow \beta + \gamma$ runs to lower temperatures.

Associated with each of the curves there exists a three-phase region made up of tie-triangles. For the $L\alpha\beta$ region, curve M_1M represents the α-phase composition, and N_1N the composition of β; for the $L\alpha\gamma$ region the α and γ compositions lie along curve M_2M and O_1O respectively; for the $L\beta\gamma$ region the β and γ compositions lie along NN_2 and OO_2 respectively.

The ternary peritectic reaction may be considered on the basis of the tie-triangles making up the three-phase regions. At a temperature just above that of the ternary peritectic reaction, the relative positions of the $L\alpha\beta$ and $L\alpha\gamma$ tie-triangles are shown in Fig. 4.9*a*. With decreasing temperature the positions of the triangles change as the compositional changes occur in the phases, and, in effect, the triangles are displaced towards one another. At the ternary peritectic temperature, the liquid composition reaches point *G*, and the α composition reaches point *M*, so that the $L\alpha$ sides of the triangles become coincident (Fig. 4.9*b*), and the two triangles form a trapezium.

At the peritectic temperature four phases *L*, α, β, and γ co-exist, the compositions of these phases lying at the corners of the trapezium *MNGO* (see Fig. 4.8). This trapezium may be termed the peritectic plane. It should be noted that the liquid composition point, *G*, lies outside the triangle formed by joining the points representing the compositions of the solid phases α, β, and γ (i.e. points *M*, *N*, and *O*); this arrangement of the phase composition points distinguishes the peritectic reaction from the invariant eutectic reaction. The peritectic reaction occurs by the interaction of *L* and α (whose compositions lie at opposite corners of the trapezium) to form a mixture of the two phases (β and γ) whose compositions lie at the other opposite corners of the trapezium.

At the completion of the invariant reaction the trapezium is divided across the $\beta\gamma$ diagonal to form two triangles which represent the phase equilibria between $\alpha\beta\gamma$ and $L\beta\gamma$ respectively (Fig. 4.9*c*). Thus, there exist two three-phase regions above the peritectic plane (namely $L\alpha\beta$ and $L\alpha\gamma$) and two below the plane (namely $\alpha\beta\gamma$ and $L\beta\gamma$).

With falling temperature the two tie-triangles $\alpha\beta\gamma$ and $L\beta\gamma$ separate from one another, as compositional changes occur in the phases (Fig. 4.9*d*). The $L\beta\gamma$ triangle is displaced as the liquid composition changes along the eutectic curve *GF*, while the β and γ compositions change along curves NN_2 and OO_2 respectively.

Alloys whose compositions lie within the $\alpha\beta\gamma$ triangle (*MNO*) at the peritectic temperature, complete their equilibrium solidification process by the ternary peritectic reaction, since all the liquid is consumed in the reaction. However, for alloys whose compositions lie in the $L\beta\gamma$ triangle (*GNO*) at the peritectic temperature, the α-phase is consumed in the invariant reaction; then with further decrease in temperature, solidification is finally completed by the displacement of the $L\beta\gamma$ triangle until the $\beta\gamma$ side of the triangle contains the alloy composition point. An alloy whose

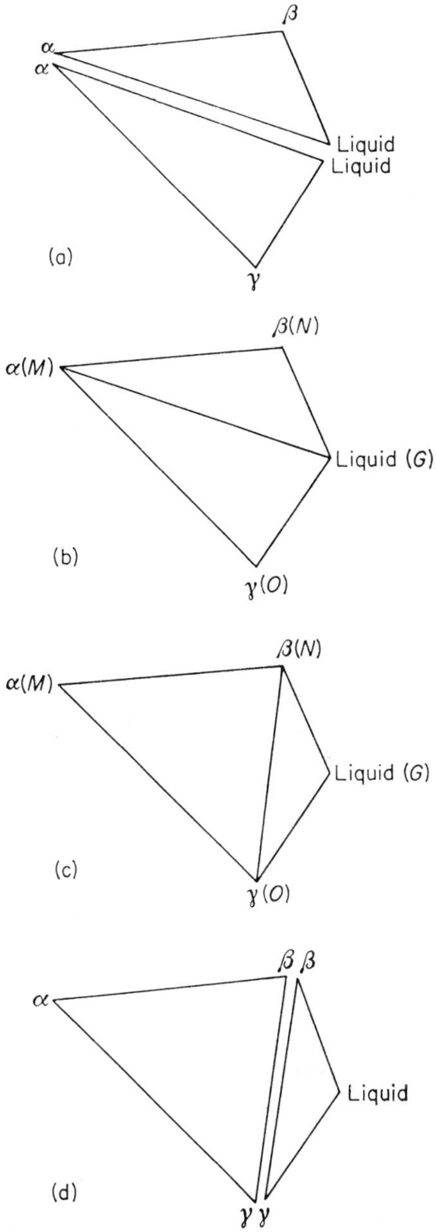

(*a*) at a temperature just above that of the ternary peritectic;

(*b*) and (*c*) at the ternary peritectic temperature;

(*d*) below the ternary peritectic temperature

Fig. 4.9. Tie-triangles

composition corresponds to the intersection of the $L\alpha$ and $\beta\gamma$ diagonals of the trapezium at the peritectic temperature is composed entirely of L and α just above this temperature, and entirely of β and γ just below it.

To illustrate the nature of the invariant reaction it is helpful to calculate

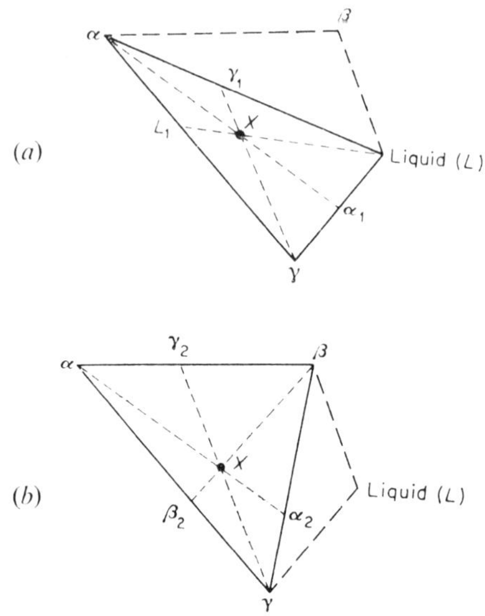

Fig. 4.10. Tie-triangles at the ternary peritectic temperature used to calculate the percentages of the phases present in alloy X before and after the occurrence of the invariant reaction

the percentages of the phases present in a typical alloy just above and below the peritectic temperature. For example, in alloy X (Fig. 4.10*a*) at the peritectic temperature at the beginning of the invariant reaction:

$$\%\ \text{Liquid} = \frac{L_1X}{LL_1} \times 100,\ \%\ \alpha = \frac{\alpha_1 X}{\alpha\alpha_1} \times 100,\ \%\ \gamma = \frac{\gamma_1 X}{\gamma\gamma_1} \times 100,$$

no β is present.

After the invariant reaction is completed (Fig. 4.10*b*):

$$\%\ \alpha = \frac{\alpha_2 X}{\alpha\alpha_2} \times 100 \text{ (i.e. less than at the beginning of the reaction),}$$

$$\%\ \gamma = \frac{\gamma_2 X}{\gamma\gamma_2} \times 100 \text{ (i.e. a greater amount than at the beginning of the reaction),}$$

$$\%\ \beta = \frac{\beta_2 X}{\beta\beta_2} \times 100,$$

no liquid is present.

Thus, during the reaction, the percentages of β and γ have increased at the expense of the L and α phases, in accordance with the peritectic nature of the process $L + \alpha \longrightarrow \beta + \gamma$.

Projected view of the system

Fig. 4.11 shows a projected view of the system. The liquidus curves are *dg*, *eg*, and *gf*, and the peritectic nature of the invariant reaction is indicated by the arrows on two of the curves pointing towards the intersection point of the curves, while the arrow on the third curve points away from the

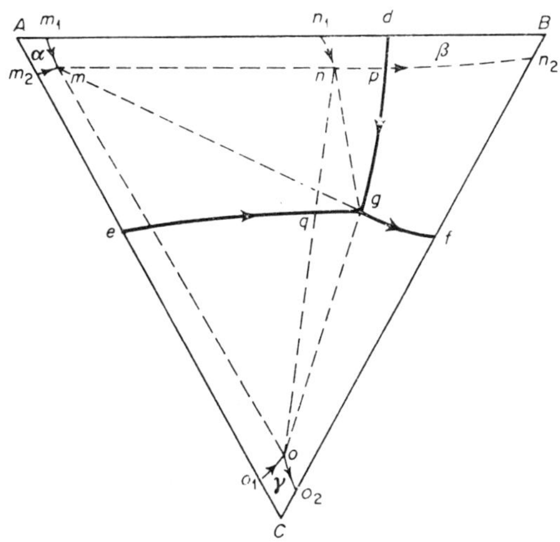

Fig. 4.11. Projected view of system of Fig. 4.8

intersection. The primary separation of α, β, and γ occurs within the regions *Adge*, *Bdgf*, and *Cegf* respectively. The solidus curves m_1m and m_2m define the limits of the α-solid solution region, while the β and γ solid solution regions are bounded by curves n_1n, nn_2 and o_1o, oo_2 respectively. The trapezium *mngo* is the projection of the isothermal plane representing the invariant reaction. As shown in Figs. 4.9*b* and 4.9*c* the upper surface of the plane is made up of the tie triangles *mng* and *mog* while the lower surface is made up of *mno* and *ngo*. All alloys whose compositions lie within the trapezium undergo the peritectic reaction.

In the solid state part of the system, the triangle *mno* (which is part of the solidus of the system) defines the three-phase $\alpha + \beta + \gamma$ region at a temperature just below the peritectic temperature. The two-phase regions $\alpha + \beta$, $\alpha + \gamma$, and $\beta + \gamma$ are m_1n_1nm, m_2moo_1, and nn_2o_2o respectively. The same types of solid state region are in fact present as in the ternary eutectic system previously discussed, but the position of the invariant liquid composition, g, relative to the positions of the α, β, and γ phases (m, n, and o respectively) is different; i.e. point g lies outside the triangle mno, not inside it as in a ternary eutectic.

The solidification of alloys

For most of the alloys lying in one of the single phase regions of Fig. 4.11 no detailed consideration of the solidification sequence is necessary here. However, it should be noted that in the region n_1dpn, solidification commences with the formation of primary α, and there follows the peritectic reaction $L + \alpha \longrightarrow \beta$ in which all the α is consumed; the liquid composition then changes along a path on the β-liquidus surface $Bdgf$, and β is deposited until solidification is completed.

With regard to the two-phase regions, alloys within the region m_1n_1nm commence their solidification by the formation of primary α, then follows the reaction $L + \alpha \longrightarrow \beta$ when the liquid composition has reached a point on curve dg; solidification is completed when the $\alpha\beta$ side of the $L\alpha\beta$ tie-triangle contains the alloy composition point. Alloys in the $\alpha + \gamma$ region (i.e. m_2moo_1) begin to solidify by depositing either α or γ, depending on the actual alloy composition; primary solidification is followed by the binary eutectic reaction $L \longrightarrow \alpha + \gamma$ that proceeds until all the liquid is consumed.

The $\beta + \gamma$ region (i.e. nn_2o_2o) is best considered in several sections:

pn_2fg. Solidification begins by the formation of primary β (i.e. Liquid (L) $\longrightarrow$ β) and then the liquid composition changes along a path on the β-liquidus until it reaches the eutectic curve gf. The reaction $L \longrightarrow \beta + \gamma$ occurs as the liquid composition changes along gf, and the β and γ compositions change along nn_2 and oo_2 respectively, until the liquid is consumed.

npg. The first stage of solidification is $L \longrightarrow \alpha$, and the liquid composition then changes until it reaches curve dg when the reaction $L + \alpha \longrightarrow \beta$ occurs. Before the liquid composition reaches point g, the α-phase will have been consumed, and the liquid composition then leaves curve dg and changes along a path on the β-liquidus while β is deposited. The liquid composition eventually reaches the eutectic curve gf when the solidification proceeds as for region pn_2fg, by the deposition of β and γ.

oo_2fg. The primary stage of solidification is $L \longrightarrow \gamma$, and the liquid composition changes on to curve gf when the $L \longrightarrow \beta + \gamma$ reaction occurs as for the region pn_2fg.

ngo. Alloys in this region undergo the invariant peritectic reaction during solidification. The primary stage of solidification is either $L \longrightarrow \alpha$ (in region *ngq*) or $L \longrightarrow \gamma$ (in region *oqg*), and the liquid composition then changes either on to the eutectic curve *eg* (i.e. $L \longrightarrow \alpha + \gamma$) or on to the peritectic curve *dg* (i.e. $L + \alpha \longrightarrow \beta$) depending on the alloy composition; with falling temperature the appropriate three-phase reaction proceeds until the liquid composition reaches the invariant peritectic point *g* when the reaction $L + \alpha \longrightarrow \beta + \gamma$ occurs. The α-phase is consumed, leaving liquid, β and γ, and with further decrease in temperature the $L \longrightarrow \beta + \gamma$ reaction proceeds along curve *gf*, the final structure being $\beta + \gamma$.

The remaining region to be considered is the three-phase region *mno*. For alloys within this region the primary separation may be either of α or of γ, and the liquid composition moves either on to curve *eg* or curve *dg*, depending on the actual alloy composition. As the three-phase reaction (i.e. either $L \longrightarrow \alpha + \gamma$ or $L + \alpha \longrightarrow \beta$) proceeds the liquid composition reaches point *g*, when the invariant reaction occurs and all the liquid is consumed under equilibrium conditions.

Isothermal sections

Fig. 4.12*a* shows an isothermal section at a temperature just above the peritectic temperature. The three-phase triangles $L + \alpha + \beta$ and $L + \alpha + \gamma$, associated with the three-phase reactions $L + \alpha \longrightarrow \beta$ and $L \longrightarrow \alpha + \gamma$ are seen in relation to the adjacent regions.

The temperature of the isothermal section shown in Fig. 4.12*b* is below that of the eutectic reaction in system BC, and is representative of the solid state constitution of the system. Assuming that no solid state solubility changes occur after solidification is complete, the limits of the phase regions in this section coincide with those shown in the projected view of the system (Fig. 4.11).

A system in which one binary system is of the eutectic type, and two are of the peritectic type, and which shows an invariant reaction of the type $L + \alpha + \beta \longrightarrow \gamma$

In the ternary system ABC illustrated in Fig. 4.13, system AB is of the eutectic type and systems BC and AC are peritectic in nature. Terminal solid solutions α, β, and γ are formed and there are no intermediate phases. (The type of invariant reaction depicted can also occur with other combinations of binary systems.[3])

The eutectic valley *DG* ($L \longrightarrow \alpha + \beta$) runs down into the ternary system. Point *G* represents the composition of liquid taking part in the invariant peritectic reaction $L + \alpha + \beta \longrightarrow \gamma$. From G, running to lower temperatures, are two peritectic curves GE ($L + \alpha \longrightarrow \gamma$) and GF ($L + \beta \longrightarrow \gamma$).

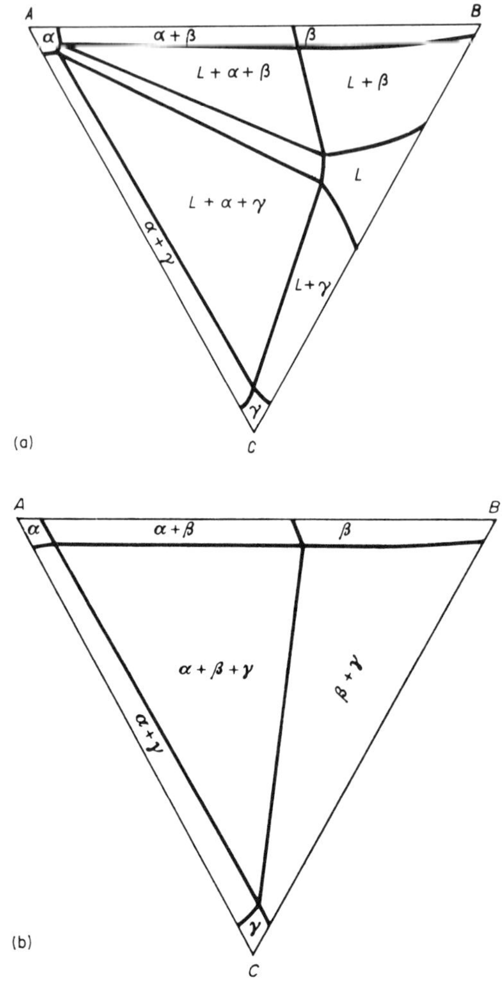

Fig. 4.12. Isothermal sections
(*a*) at a temperature just above that of the ternary peritectic
(*b*) at a temperature below that of the eutectic reaction in system *BC*

For the eutectic reaction $L \longrightarrow \alpha + \beta$, as the liquid composition changes along DG, the α and β compositions change along M_1M and N_1N respectively. For the peritectic reaction $L + \alpha \longrightarrow \gamma$, the compositions of the liquid, α and γ change along the curves *GE*, MM_2, and OO_1 respectively; while for the $L + \beta \longrightarrow \gamma$ reaction, the compositional changes of liquid, β and γ occur along *GF*, NN_2, and OO_2 respectively.

The essential nature of the invariant reaction may be considered on the basis of the tie-triangles that constitute the three-phase regions in the

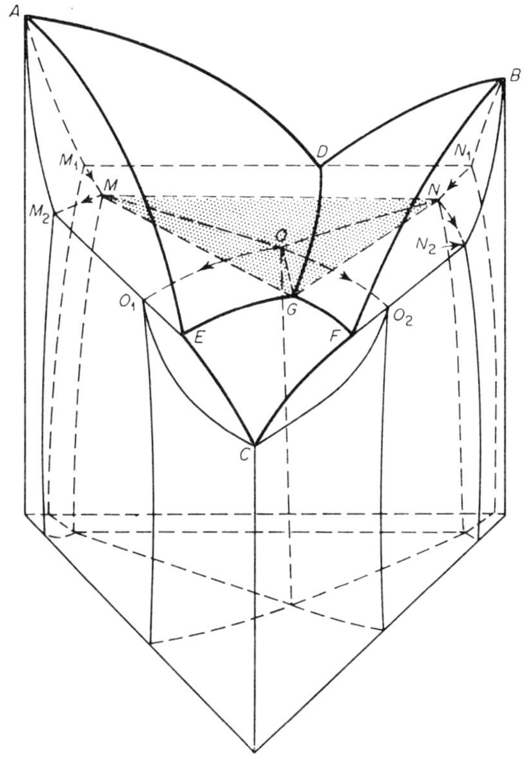

Fig. 4.13. Space model of system showing a ternary peritectic reaction $L + \alpha + \beta \longrightarrow \gamma$

system. One three-phase region, namely $L + \alpha + \beta$, extends down on to the plane that represents the invariant reaction. A typical $L\alpha\beta$ tie-triangle through this space, at a temperature just above the peritectic, is shown in Fig. 4.14*a*. At the invariant reaction temperature, this triangle (*GMN*) defines the compositions of the liquid, α and β that take part in the reaction. The composition of the γ-phase lies at point *O*, within the triangle *GMN*; thus the liquid composition *G* lies outside the triangle (*MNO*) formed by the compositions of the three solids, which is a characteristic feature of the peritectic reaction. The reaction gives rise to three three-phase regions below the peritectic plane, represented by the tie-triangles $\alpha\beta\gamma$ (*MNO*), $L\alpha\gamma$ (*GMO*), and $L\beta\gamma$ (*GNO*) in Fig. 4.14*b*. With decreasing temperature these three tie-triangles become separated as compositional changes occur (Fig. 4.14*c*); the triangles $L\alpha\gamma$ and $L\beta\gamma$ are associated with the reaction $L + \alpha \longrightarrow \gamma$ and $L + \beta \longrightarrow \gamma$ respectively.

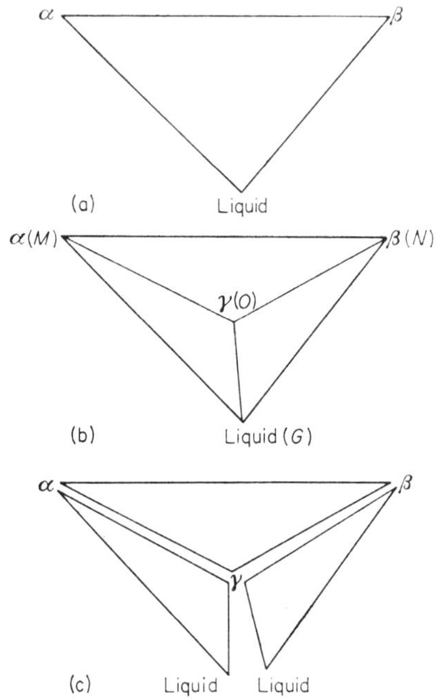

Fig. 4.14. Tie-triangles
(*a*) at a temperature just above that of the ternary peritectic
(*b*) at the peritectic temperature
(*c*) below the peritectic temperature

Alloys whose compositions lie within the $\alpha\beta\gamma$ region at the peritectic temperature complete their solidification by the invariant reaction, all the liquid being consumed. However, if the alloy composition lies either in the $L\alpha\gamma$ or $L\beta\gamma$ region of the peritectic plane, then some liquid remains at the completion of the invariant reaction, and the solidification sequence continues via either the $L + \alpha \longrightarrow \gamma$ or the $L + \beta \longrightarrow \gamma$ reaction.

Projected view of the system

Fig. 4.15 shows a projected view of the system. The liquidus curves are *dg*, *ge*, and *gf*, and the type of peritectic reaction is indicated by the fact that the arrow on the first of these curves points to the intersection of the curves, and the arrows on the other two curves point away from the intersection. The primary separations of α, β, and γ occur within the regions *Adge*, *Bdgf*, and *Cegf* respectively. The limits of the α, β, and γ regions are defined by the solidus curves m_1m, mm_2; n_1n, nn_2, and oo_2, oo_1 respectively. Triangle *mng* is the projection of the isothermal reaction

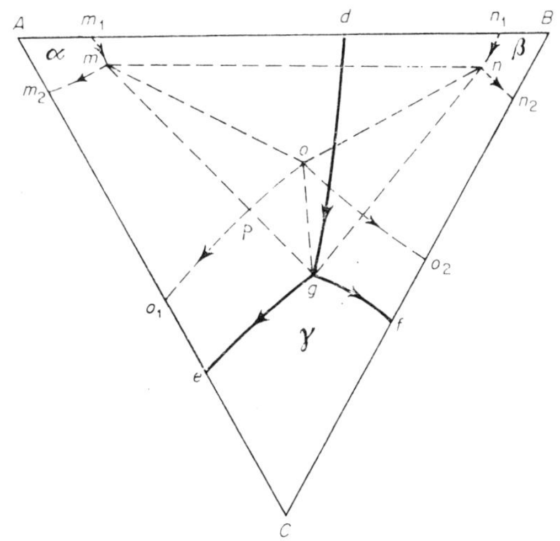

Fig. 4.15. Projected view of system of Fig. 4.13

plane, and all alloys whose compositions lie within this triangle undergo the peritectic reaction. The division of the triangle on the underside into the tie-triangles *mno*, *mog*, *ngo* is also shown. Triangle *mno* is part of the solidus of the system and in the solid state part of the system represents the three-phase αβγ region. The two-phase regions α + β, α + γ, and β + γ are m_1n_1nm, m_2moo_1, and nn_2o_2o respectively.

The solidification of alloys

The solidification sequences for alloys that undergo the peritectic reaction (i.e. alloys within triangle *mng*) illustrate the main features of the system. Following the primary separation of α or β, the liquid composition moves on to the curve *dg*, and then along this curve as the reaction $L \longrightarrow \alpha + \beta$ proceeds until the invariant point *g* is reached, when the reaction $L + \alpha + \beta \longrightarrow \gamma$ occurs. If the alloy composition lies within triangle *mno* all the liquid is consumed in the reaction.

For alloys in the region *mog* all the β is consumed, leaving liquid, α and γ, and with falling temperature the peritectic reaction $L + \alpha \longrightarrow \gamma$ proceeds, the liquid composition changing along *ge*, and the α and γ compositions along mm_2 and oo_1 respectively. If the alloy composition is within region *mop*, so that the alloy consists of $\alpha + \gamma$ in the solid state, all the liquid is consumed in the reaction $L + \alpha \longrightarrow \gamma$. For alloy compositions within region *ogp*, which are single phase γ when solid, all the α is consumed in the $L + \alpha \longrightarrow \gamma$ reaction and the remaining liquid moves away from curve *ge*, over the γ liquidus surface, while γ is deposited until solidification is completed.

For alloys in the region *ngo*, the α is consumed in the invariant reaction, leaving liquid, β and γ, and solidification then continues via the $L + \beta \longrightarrow \gamma$ reaction along curve *gf*, in a manner analogous to that for region *mog*.

The solidification of alloys in the remaining regions of the system is not discussed here; the general principles previously discussed for other systems are applicable.

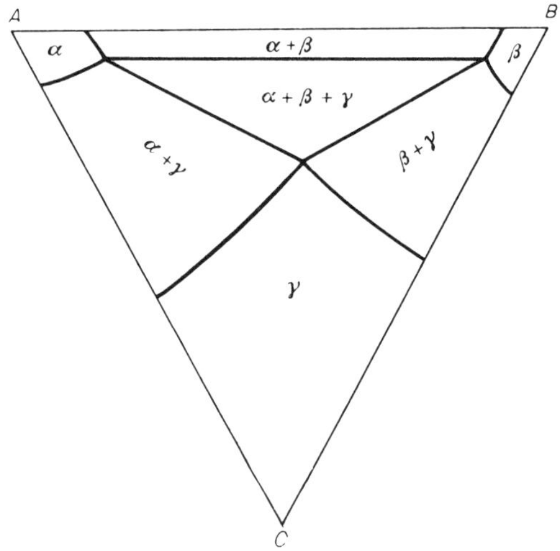

Fig. 4.16. Isothermal section in the solid state for system of Fig. 4.13

Isothermal section

Fig. 4.16 shows an isothermal section in the solid state, and if solubility changes following solidification are neglected, the limits of the phase regions coincide with those on the projected view (Fig. 4.15).

Problems

1. With reference to Fig 4.1:

(a) State what phase regions are enclosed by the following surfaces:

(i) BDN_1, $BDGF$, DN_1NG, FN_2NG, BN_1NN_2, BFN_2; (ii) M_2MGE, O_1OGE, M_2MOO_1, MGO.

(b) State what surfaces separate the following phase regions: (i) $L + \alpha + \beta/L + \alpha$; (ii) $\gamma/L + \gamma$.

2. With reference to Fig. 4.8:

(a) State what surfaces enclose the following phase regions: (i) $L + \alpha + \beta$, (ii) $L + \alpha$.

(b) State what phase regions are separated by the surfaces: (i) $OGFO_2$; (ii) MNO; (iii) NGO.

3. The melting points of metals, A, B and C are 1000°C, 950°C and 850°C, respectively. The metals show complete liquid solubility, but negligible solid solubility in one another: they form no compounds.

The following table gives a complete list of the temperatures of cooling curve discontinuities and arrests for certain alloys:

Alloy number	Alloy composition wt%A	wt%B	wt%C	Temperatures (°C)
1	60	40	—	750
2	51	—	49	800
3	—	44	56	700
4	80	10	10	950 (primary A); 600
5	30	50	20	850 (primary B); 600

(a) Draw a liquidus projection for the ternary system ABC that conforms with the above data.

(b) Calculate the proportion of primary A that forms during the solidification of the alloy containing 80% A, 10% B and 10% C.

4. Ag–Cu–Cu_3P. Part of the Cu–Ag–P system is represented by the liquidus projection in Fig. 4.17, taking the compound Cu_3P as one of the components. Each of the binary systems contains a eutectic and the ternary system contains an invariant eutectic $L \leftrightharpoons Ag + Cu + Cu_3P$. (Solid solubility, which is small at room temperature, is neglected here.)

The system includes a group of brazing alloys (in the range 0–15 wt% Ag; 40–60 wt% Cu; 35–55 wt% Cu_3P) used for joining copper alloys.

(a) For an alloy containing 50 wt% Cu_3P and 5 wt% Ag determine: (i) the liquidus temperature; (ii) the percentage of liquid present at the temperature when separation of Cu_3P begins; (iii) the proportions of the

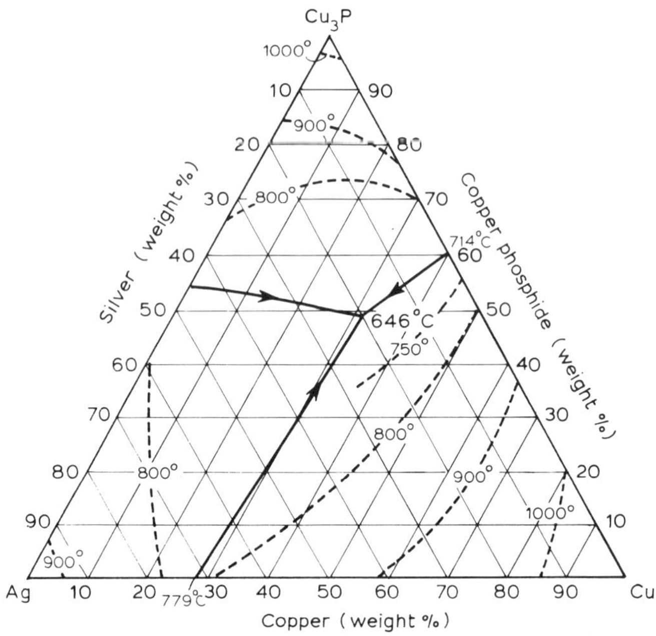

Fig. 4.17. Ag–Cu–Cu_3P system. From G. H. Sistaire (1973) *Metals Handbook* (8th edition), American Society for Metals, **8**, p. 379

phases present at the stage when the liquid contains 10 wt % Ag and lies on the $L \longrightarrow Cu + Cu_3P$ eutectic valley; and (iv) the percentages of primary phase and of binary and ternary eutectic mixtures respectively present at room temperature. (Assume equilibrium conditions and neglect solid solubility.)

(b) Sketch isothermal sections through the system at (i) 800°C and (ii) at a temperature above 646°C but below the binary eutectic temperatures. (Make assumptions as necessary concerning the liquidus isotherms and neglect solid solubility.)

5. Three metals A, B and C show complete liquid solubility in one another and partial solid solubility. The ternary system contains only three solid phases namely α, β and γ solid solutions. The melting points of A, B and C are 900°C, 1200°C and 1000°C, respectively.

The system contains invariant reactions at the following temperatures and compositions:

Temperature (°C)	Composition (wt %)		
	A	B	C
1100	—	35	65
1000	70	30	—
950	45	20	35
800	50	—	50

Solid solubility data are as follows:

Solid Solution	Solid solubility limits (wt %)	
	at 945°C	at room temperature
α	50 A, 40 B, 10 C	60 A, 35 B, 5 C
β	12 A, 80 B, 8 C	3 A, 95 B, 2 C
γ	10 A, 45 B, 45 C	5 A, 40 B, 55 C

(a) Draw a liquidus projection of the system and an isothermal section at room temperature, consistent with the above data.

(b) State the nature of each invariant reaction in the system.

(c) Describe the equilibrium solidification sequence of an alloy containing 30% A, 55% B, 15% C, and calculate the relative amounts of the phases present in this alloy at 945°C.

(d) Describe the equilibrium solidification sequence of an alloy containing 20% A, 35% B, 45% C.

5
Systems Containing more than Four Phases

THE presence of one or more intermediate phases (either binary or ternary) in addition to three primary solid solutions (or pure components) leads to many possible forms of ternary system. Several examples are discussed, but the space models depict only the liquidus features and not the phase regions within the models.

An intermediate phase stable up to its melting point (i.e. congruently melting) exists in one of the binary systems

In the system ABC depicted in Fig. 5.1, it is assumed that system AB contains an intermediate phase, X, that is stable up to its melting point, while systems AC and BC are eutectic in nature. Four binary eutectic curves enter the ternary system from the binary systems, and assuming that there is no solid solubility of the components, the reactions are as follows:

$$\begin{aligned} &\text{DI}: L \longrightarrow A + X \\ &\text{EG}: L \longrightarrow B + X \\ &\text{JI}: L \longrightarrow A + C \\ &\text{FG}: L \longrightarrow B + C \end{aligned}$$

Within the ternary system there is also a curve IHG representing the eutectic reaction $L \longrightarrow X + C$. As shown in Fig. 5.1 this curve is in the form of a 'saddle' with a maximum point at H.

The system contains two invariant eutectic reactions at points I and G:

$$\begin{aligned} &\text{At I}: L \longrightarrow A + X + C \\ &\text{At G}: L \longrightarrow B + X + C \end{aligned}$$

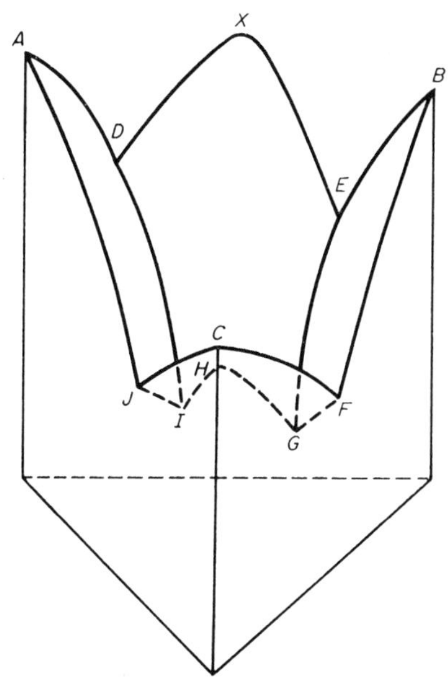

Fig. 5.1. Space model of system containing an intermediate phase X, stable up to its melting point, and showing two invariant reactions $L \longrightarrow A + X + C$ and $L \longrightarrow B + X + C$ (Only liquidus features are shown)

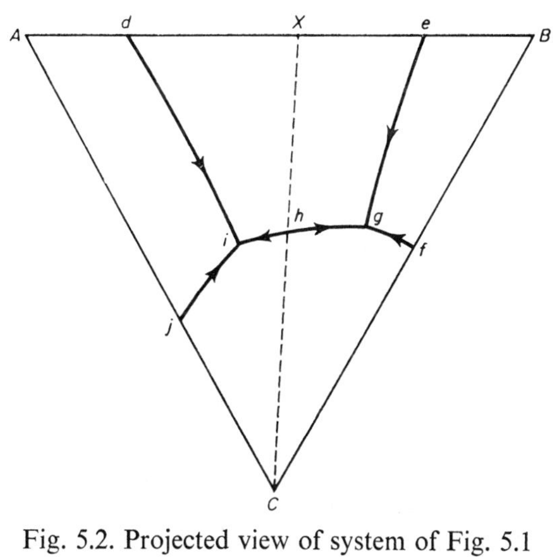

Fig. 5.2. Projected view of system of Fig. 5.1

The regions of primary solidification of *A*, *B*, *C*, and *X* shown in the projected view of Fig. 5.2, are *Adij*, *Begf*, *Cjihgf*, and *deghi* respectively. The line *XC* divides the system into two triangular portions *AXC* and

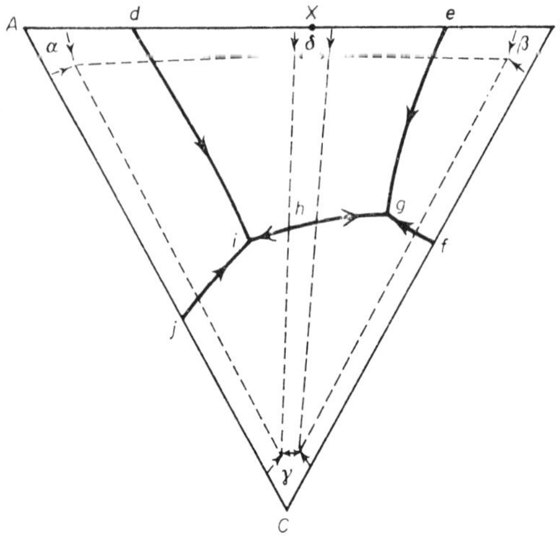

Fig. 5.3. Projected view of system of Fig. 5.1 assuming partial solid solubility of the components

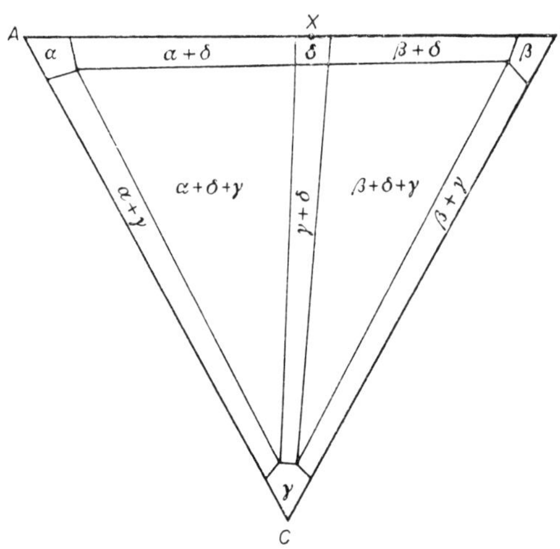

Fig. 5.4. Isothermal section in the solid state for the system of Fig. 5.1 showing partial solid solubility

BXC, each of which is a ternary eutectic type of system with no solid solubility of the components: the components in these two portions are A, X, and C, and B, X, and C respectively.

XC is termed a 'quasi-binary' section or system; it has the characteristics of a binary system in that during the course of solidification of any alloy lying in it, the compositions of liquid and solid remain in the section, i.e. all tie-lines, and the three-phase equilibrium line all lie in the plane of the section.

The form of section XC (shown in Fig. 5.5) is eutectic in type, and the eutectic point for the reaction $L \longrightarrow X + C$, corresponds to the saddle-point, H, in Fig. 5.1.

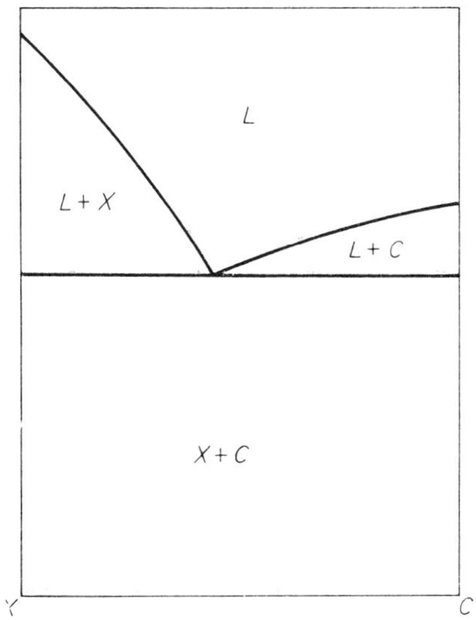

Fig. 5.5. Quasi-binary section XC in system of Fig. 5.1

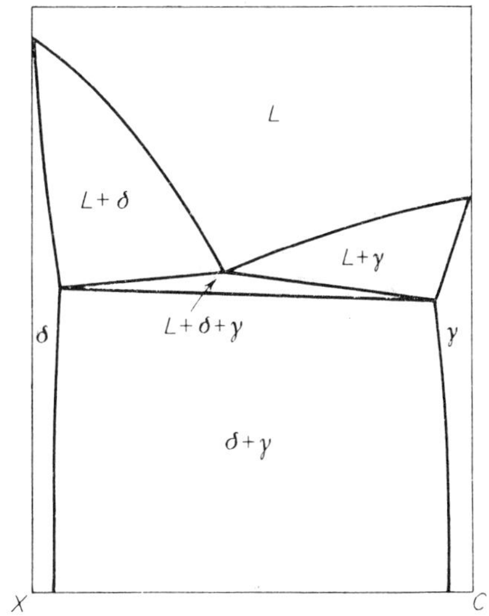

Fig. 5.6. Section XC (not strictly quasi-binary), when system shows partial solid solubility

If the ternary system is assumed to show partial solid solubility of the components, it may be represented in a projected view as in Fig. 5.3, while an isothermal section in the solid state is shown in Fig. 5.4.

When there is partial solid solubility as illustrated, the section joining C to the maximum point X in the binary system AB is not necessarily quasi-binary in the exact sense defined above. For example, the section might be as shown in Fig. 5.6, since the 'saddle top' of the $L + \delta + \gamma$ region does not necessarily lie in the plane XC.[6] However, in practice, the term quasi-binary is commonly applied to cases where some degree of solid solubility exists, and if the extent of the solubility is small there is likely to be little deviation from the strictly quasi-binary case.

The solidification sequences of typical ternary alloys will not be discussed here; the principles previously described for ternary eutectic systems are applicable.

With the same types of binary system as in the case described above, it is possible for the ternary system to contain one invariant peritectic reaction and one invariant eutectic reaction, instead of containing two invariant eutectics. This alternative case is illustrated in Fig. 5.7 and in

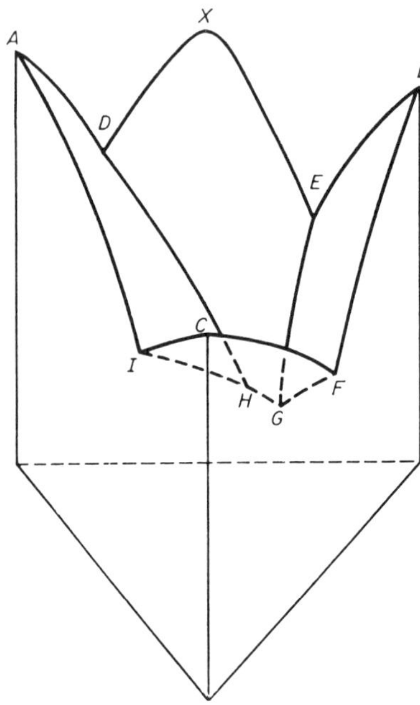

Fig. 5.7. Space model of system containing intermediate phase X, and showing two invariant reactions $L + A \longrightarrow X + C$ and $L \longrightarrow B + X + C$. (Only liquidus features are shown)

the projected view shown in Fig. 5.8; the regions of primary solidification of A, B, C, and X are *Adhi*, *Begf*, *Cihgf*, and *degh* respectively. Assuming that there is no solid solubility, point H represents the composition of liquid taking part in the reaction

$$L + A \longrightarrow X + C$$

The peritectic reaction arises because the intersection of the curves $DH(L \longrightarrow A + X)$ and $IH(L \longrightarrow A + C)$ occurs at a point H which lies outside the triangle formed by the compositions of the solid phases A, X, and C. Curve HG representing the reaction $L \longrightarrow X + C$ runs to lower temperatures from H, and meets the eutectic curves $EG(L \longrightarrow B + X)$ and $FG(L \longrightarrow B + C)$ to give rise to the invariant reaction at $G(L \longrightarrow B + X + C)$.

In the solid state, the system is divided into the regions AXC and BXC as in the first case discussed. However, section XC is not quasi-binary; consideration of the solidification of alloys lying in this section will show that the compositions of the liquid and solids do not remain in the plane of the section throughout the whole course of solidification, although the final structure under equilibrium conditions consists of X and C.

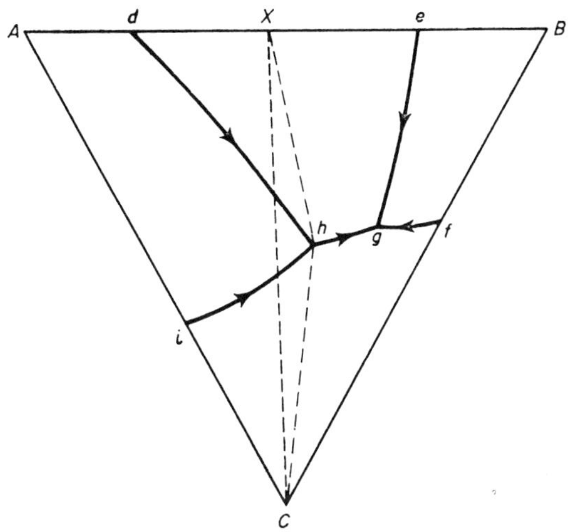

Fig. 5.8. Projected view of system of Fig. 5.7

All alloys lying within the region $AXhC$ (Fig. 5.8) undergo the ternary peritectic reaction during solidification, the liquid compositions reaching point h, either via curve ih or curve dh. For alloys in the triangle AXC, all the liquid is consumed in the reaction under equilibrium conditions. If the alloy composition lies in the triangle XhC, all of component A is consumed, and the remaining liquid deposits X and C by a eutectic reaction along curve hg as cooling proceeds, until the liquid composition reaches g when the reaction $L \longrightarrow B + X + C$ completes the solidification process. If the alloy composition lies on the line XC then both liquid and A are consumed in the peritectic reaction leaving a mixture of X and C.

Fig. 5.9 shows a projected view of the system assuming that partial solid solubility exists.

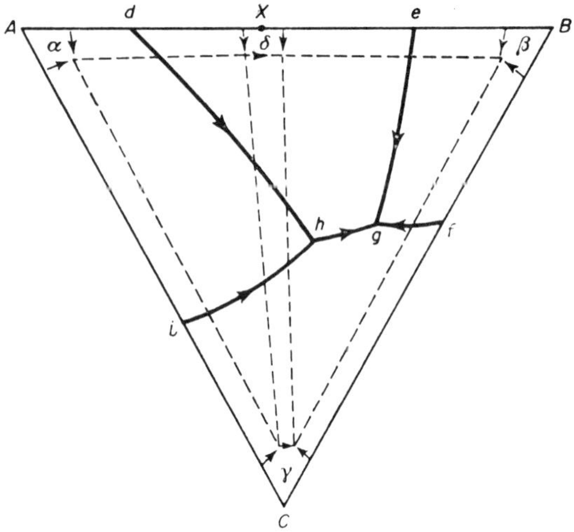

Fig. 5.9. Projected view of system of Fig. 5.7, assuming partial solid solubility of the components

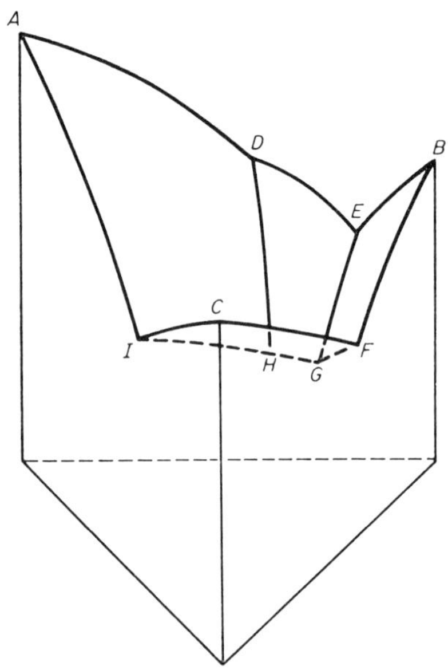

Fig. 5.10. Space model of system containing an intermediate phase X (formed by a peritectic reaction), and containing two invariant reactions $L + A \longrightarrow X + C$ and $L \longrightarrow B + X + C$. (Only liquidus features are shown)

An intermediate phase formed by a peritectic reaction exists in one of the binary systems

In the system ABC shown in Figs. 5.10 and 5.11, in which it is assumed that there is no solid solubility of the components, systems AC and BC

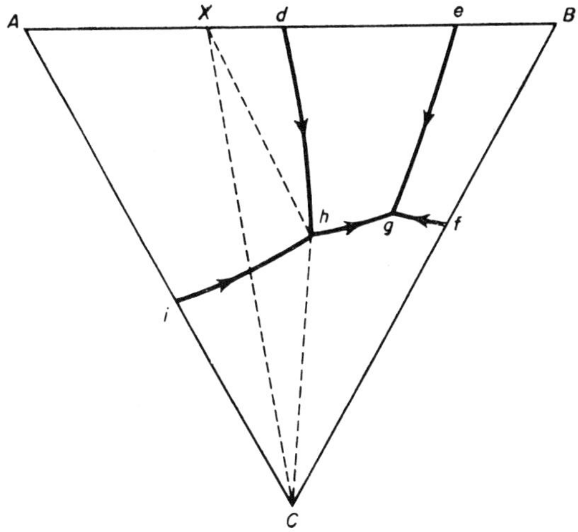

Fig. 5.11. Projected view of system of Fig. 5.10

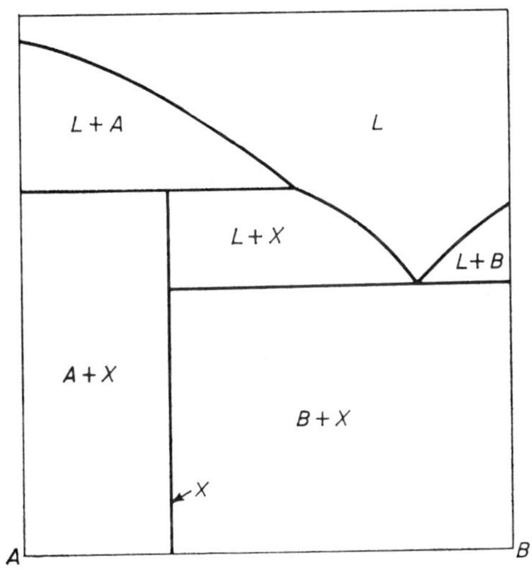

Fig. 5.12. Binary system AB of ternary system shown in Fig. 5.10

are eutectic in nature, while system AB contains an intermediate phase, X, formed by the reaction $L + A \longrightarrow X$ (see Fig. 5.12); in the ternary system the peritectic curve $DH(L + A \longrightarrow X)$ intersects the eutectic curve $IH(L \longrightarrow A + C)$ at the invariant peritectic point $H(L + A \longrightarrow X + C)$. Curve HG representing the reaction $L \longrightarrow X + C$ intersects the eutectic curves $EG(L \longrightarrow X + B)$ and $FG(L \longrightarrow B + C)$ to give the reaction $L \longrightarrow B + X + C$ at point G. The portion of the liquidus surface $DEGH$ represents the primary separation of the intermediate phase X.

In the solid state the system is divided into the two regions AXC and BXC as in the case previously discussed, and section XC is not quasi-binary. All alloys within the region $AXhC$ (Fig. 5.11) undergo the peritectic reaction and, depending on whether an alloy lies within the triangle AXC

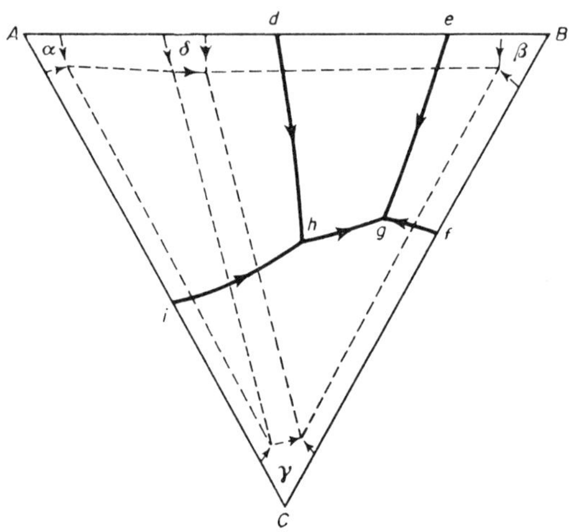

Fig. 5.13. Projected view of system of Fig. 5.10 assuming partial solid solubility

or XhC, either the liquid or component A is consumed, as discussed for the previous case. The solidification sequence of alloys in the region Xdh is as follows: the primary separation of A causes the liquid composition to reach curve dh when the reaction $L + A \longrightarrow X$ occurs; all the A is consumed and the liquid composition then leaves curve dh and changes on a path on the surface $degh$ as X is deposited, until the liquid composition reaches either hg or eg, and then finally attains point g.

Fig. 5.13 shows a projected view of the system assuming partial solid solubility of the components.

It is possible that the curve dh representing the $L + \alpha \longrightarrow \delta$ peritectic reaction may change direction as it moves into the ternary system, and thus undergo a transition to a eutectic reaction, $L \longrightarrow \alpha + \delta$, as discussed in a previous section. In this case the junction of the $L \longrightarrow \alpha + \delta$ curve with the $L \longrightarrow \alpha + \gamma$ curve, *ih*, within the triangle formed by the compositions of the three solids α, δ, and γ leads to an invariant eutectic reaction $L \longrightarrow \alpha + \delta + \gamma$. The system then contains two invariant eutectics,

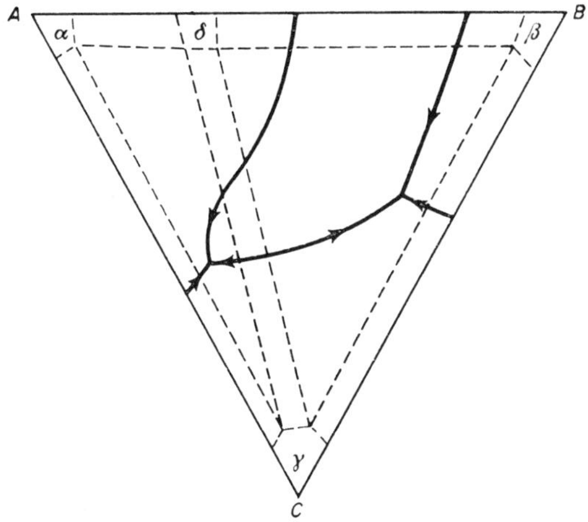

Fig. 5.14. Projected view of system in which the binary peritectic reaction $L + \alpha \longrightarrow \delta$ changes to a eutectic reaction $L \longrightarrow \alpha + \delta$

connected by a saddle-type eutectic curve $L \longrightarrow \delta + \gamma$ (Fig. 5.14), but section *XC* is not quasi-binary (see also Fig. 5.15[9] and page 132).

More than one binary intermediate phase exists in the system

Figures 5.16*a* and *b* illustrate two possible arrangements of quasi-binary sections in a ternary system ABC in which systems AB and BC each contain a congruently-melting intermediate phase, designated X and Y respectively.

In Fig. 5.16*a*, *XY* and *XC* represent quasi-binary sections, while in Fig. 5.16*b*, *XY* and *YA* are quasi-binary. In both cases, the ternary system is divided into three parts, each of which may be considered as an independent ternary system. It follows from the nature of quasi-binary sections that they cannot cross one another, and thus it is not possible for *XC* and *YA* to be the pair of quasi-binary sections.

To determine experimentally the particular pair of quasi-binary sections in an actual system (if in fact such exist), an alloy may be made at the intersection of the lines *XC* and *YA*, and after being annealed to equilibrium, it may be examined to find what phases are present. If only phases

X and C are found, then XC is quasi-binary, and if A and Y are found, then *YA* is the quasi-binary section, assuming that peritectic reactions are not involved.

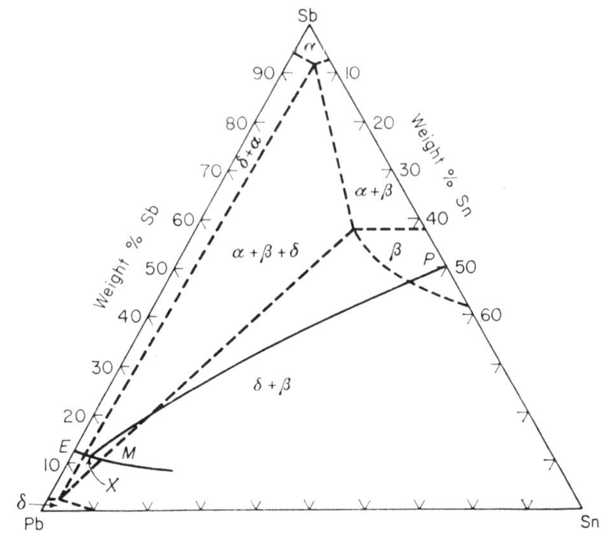

Fig. 5.15. Liquidus projection (shown as full lines) of part of the Pb–Sb–Sn system, together with part of a solid state isothermal section (dashed lines) for the system. α and δ are primary solid solutions based on Sb and Pb respectively, and β represents the solid solution based on the intermediate phase, SbSn. The reactions represented on the liquidus projection are as follows:

EX: $L \longrightarrow \alpha + \delta$
MX: $L \longrightarrow \beta + \delta$
PX: $L + \alpha \longrightarrow \beta$
X: $L \longrightarrow \alpha + \beta + \delta$

(From *Metals Handbook*, 7th Edn., 1948)

Ternary intermediate phases are present

No detailed consideration is given here of systems containing one or more ternary intermediate phases. A hypothetical example is illustrated in Fig. 5.17, which represents a system containing a congruently-melting ternary phase, Z, the binary systems each being of the eutectic type. Three quasi-binary sections, *AZ*, *BZ*, and *CZ* exist, and the system is divided into three parts.

Rules relating the maximum number of independent ternary systems into which a ternary system can be divided when binary and/or ternary intermediate phases occur are discussed and illustrated by Rhines,[7] who also illustrates typical systems.

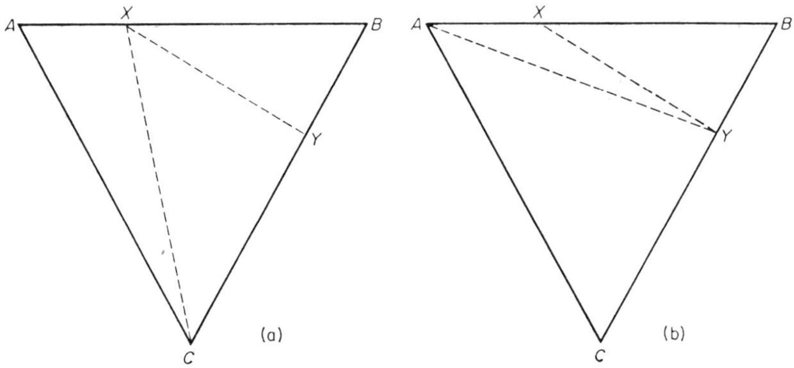

Fig. 5.16. Possible arrangements of quasi-binary sections in a system containing two binary intermediate phases

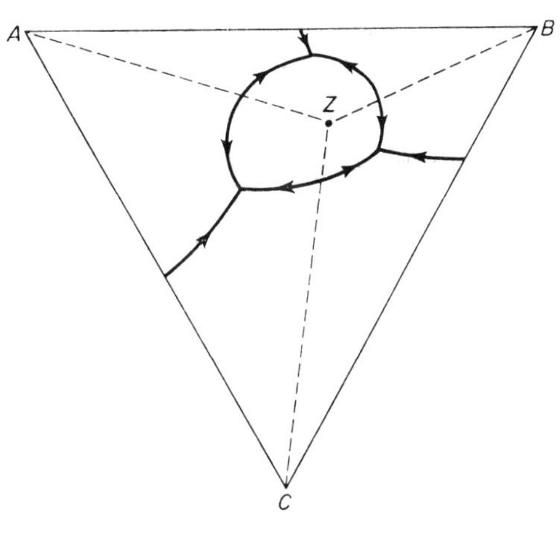

Fig. 5.17. Projected view of a system containing a ternary intermediate phase

Examples of actual systems containing both binary and ternary intermediate phases are common, e.g. in metallic and ceramic systems. Fig. 5.18 *a* and *b*, illustrate part of the Al–Cu–Mg system[12] and Fig. 5.19 *a* and *b* depict a complex oxide system.[8]

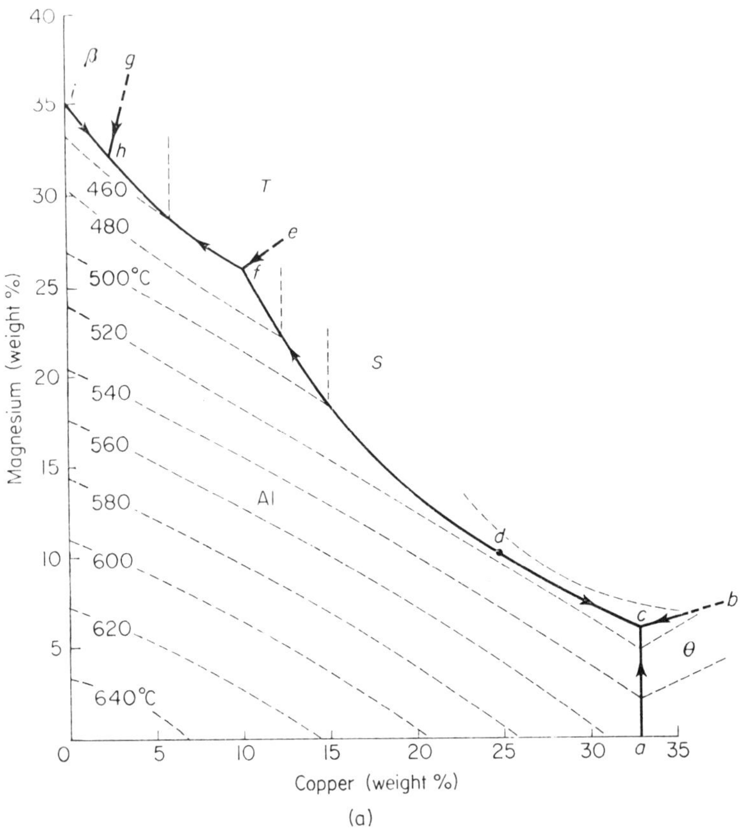

Fig. 5.18 (*a*). Liquidus projection of part of the Al–Cu–Mg system. The phases involved are as follows: aluminium solid solution (Al); $CuAl_2(\theta)$; $Mg_2Al_3(\beta)$; a ternary intermediate phase corresponding approximately to Al_2CuMg (S); and a ternary intermediate phase corresponding approximately to Al_6CuMg_4 (T). The fields of primary phase separation are labelled and some isotherms are shown. The reactions represented are as follows:

ac: $L \longrightarrow \text{Al} + \theta$	f: $L + \text{S} \longrightarrow \text{Al} + \text{T}$
bc: $L \longrightarrow \theta + \text{S}$	fh: $L \longrightarrow \text{Al} + \text{T}$
c: $L \longrightarrow \text{Al} + \theta + \text{S}$	gh: $L \longrightarrow \beta + \text{T}$
cdf: $L \longrightarrow \text{Al} + \text{S}$	ih: $L \longrightarrow \text{Al} + \beta$
ef: $L + \text{S} \longrightarrow \text{T}$	h: $L \longrightarrow \text{Al} + \beta + \text{T}$

The section Al–S is approximately quasi-binary, and *d* is the saddle point on curve *cf*.

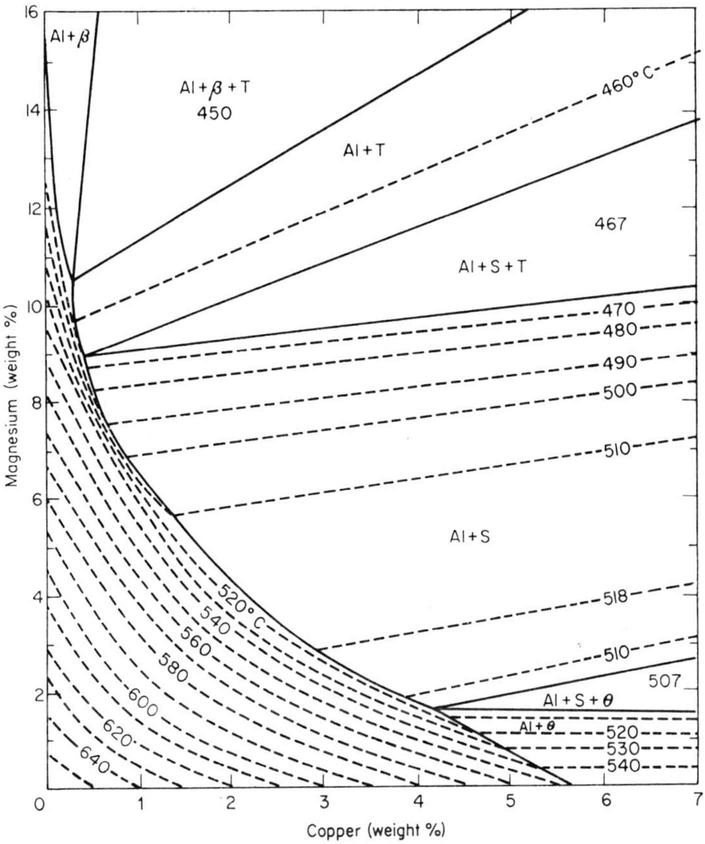

(b)

Fig. 5.18 *b*. Solidus projection of part of Al–Cu–Mg system[12]

Reproduced from Information Bulletin 25, *Equilibrium Diagrams of Aluminium Alloy Systems*, available from the Aluminium Federation

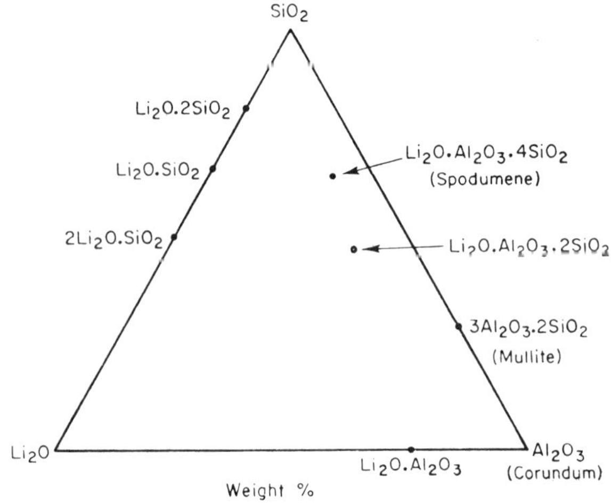

(a)

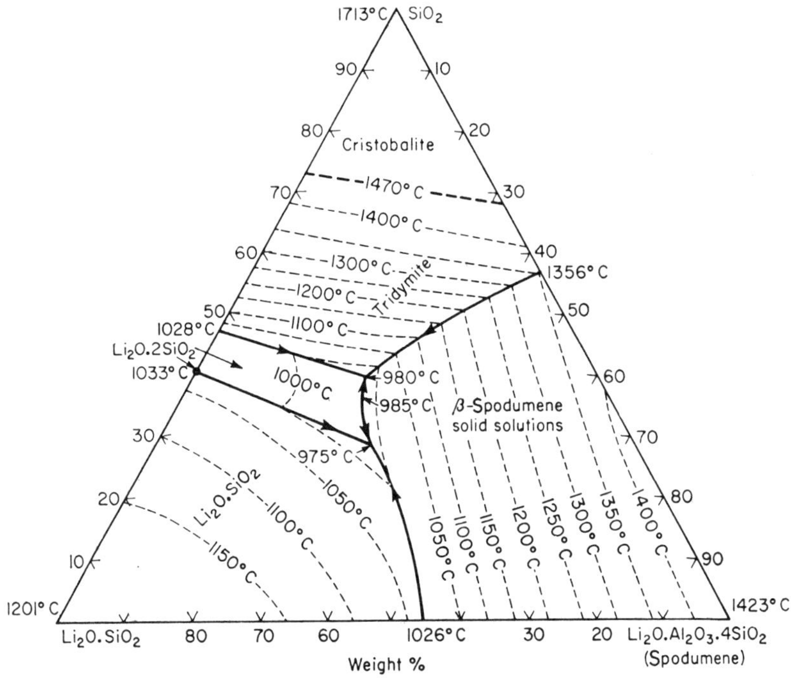

(b)

Fig. 5.19 (*a*). Diagram showing the binary and ternary phases present in the ternary system Li_2O–SiO_2–Al_2O_3

(*b*) Liquidus projection for the part of the above system SiO_2–$Li_2O.SiO_2$–$Li_2O.Al_2O_3.4SiO_2$, the composition scales being plotted taking these phases as the components. (From *Phase Diagrams for Ceramists*, The American Ceramic Society Inc., 1956.)

Problems

1. A ternary system ABC shows complete liquid solubility and partial solid solubility, forming primary solid solutions α, β and γ, based on A, B and C, respectively. System BC contains a peritectic reaction (liquid + $\beta \rightleftharpoons \delta$) at 800°C, in which δ is an intermediate phase containing 40% C, and the liquid contains 50% C.

The ternary system contains two invariant reactions as follows:

Temperature	Liquid composition	Solid phases involved
650°C	30% A, 32% B, 38% C	α, β, δ
620°C	37% A, 8% B, 55% C	α, δ, γ

The solid solubility ranges of α, β, γ and δ are very small (< 1%).

(a) Draw: (i) a liquidus projection for the system consistent with the data, and state the nature of the reactions shown; (ii) an isothermal section corresponding to a temperature of 600°C.

(b) By reference to the liquidus projection and the 600°C isothermal describe a possible solidification sequence for an alloy containing 10% A, 60% B and 30% C, assuming equilibrium cooling conditions. Calculate the equilibrium proportions of β phase present in this alloy at 651°C and 649°C, respectively.

2. With reference to Fig. 5.15 describe the equilibrium solidification sequences of the following alloys: (a) Pb–15% Sb–5% Sn; (b) Pb–15% Sb–10% Sn.

3. (a) With reference to Figs 5.18*a* and *b* describe the equilibrium solidification sequences of the following Al-based alloys: (i) 3% Cu, 15% Mg; (ii) 6% Cu, 12% Mg; (iii) 24% Cu, 10% Mg; (iv) 4.5% Cu, 2.5% Mg.

(b) Samples of an Al–4.5% Cu–2.5% Mg alloy and of a binary Al–4.5% Cu alloy are solution treated at 500°C for several hours, water-quenched and then heated at 200°C for 24 hours. From the data in Fig. 5.18*b* and taking the solubilities of Cu in Al and Mg in Al at 200°C as ~0.1 wt% and ~3 wt%, respectively, comment on the difference in the heat treatment response of these two alloys.

4. With reference to the system SiO_2–$Li_2O \cdot SiO_2$–$Li_2O \cdot Al_2O_3 \cdot 4SiO_2$:

(a) Draw an isothermal section for a temperature of 970°C. Make use of the following information in addition to that shown in Fig. 5.19. In the spodumene–SiO_2 binary system, β-spodumene shows extensive solid solubility for SiO_2. β-spodumene co-exists with $Li_2O \cdot SiO_2$ over the range

0–23% SiO_2 and with $Li_2O{\cdot}2\ SiO_2$ over the range 23–33% SiO_2. Assume negligible solid solubility in the following compounds: $Li_2O{\cdot}SiO_2$, $Li_2O{\cdot}2SiO_2$ and SiO_2.

(b) Draw a graph showing how the percentage of liquid present in equilibrium at 1100°C varies with composition along a line joining SiO_2 to a binary composition of 20% spodumene–80% $Li_2O{\cdot}SiO_2$.

6
Principles of Using Liquidus Projections

EQUILIBRIUM data presented as liquidus projections are often used to study solidification sequences. In order to make the fullest use of such projections, information is also required on the solid state constitution of a given system at the completion of solidification.

Two ways of presenting the data on solid state constitution may be considered, namely, a solidus projection, and an isothermal section at a temperature below the lowest solidus temperature in the system. In general, in systems with no solid solubility (e.g. as is effectively the case in many ceramic systems), these two modes of presentation are equivalent for the purpose required. In such systems the solidification of all ternary alloys is completed by an invariant reaction, and the solidus is composed of a plane, or planes, depending on the number of invariant reactions contained in the system. The solidus projection and the isothermal section show one or more three-phase regions, bounded by lines joining pairs of phases co-existing in equilibrium. In systems with solid solubility, the solidus projection is the most accurate and informative means of presenting the required data, especially if isotherms are depicted. If significant solid solubility changes or phase transformations occur in the solid state, the isothermal section at a temperature just below the lowest solidus temperature may not be a satisfactory source of information for studying the solidification sequences of various alloys. Further, in such cases, isothermal sections at various temperatures through the solid state region would be needed if the changes occurring during the cooling of given alloys to low temperatures are to be traced accurately.

The preceding chapters have included accounts of the procedures involved in studying solidification sequences by reference to liquidus and solidus projections: various examples have been considered, some of them relatively complex. In this chapter, a summary is given of the basic principles and procedures; it is hoped that this will assist the reader in reviewing the examples previously discussed and in the consideration of other systems.

Primary phase fields on a liquidus projection

The liquidus projection for a system is commonly labelled to show the regions of primary solidification, and the location of a given alloy composition within a particular phase field on the projection serves to indicate the primary phase that separates during solidification. If isotherms are shown on the liquidus projection, they can be used to determine the liquidus temperature for a given alloy. As solidification proceeds by separation of primary phase the composition of the liquid changes progressively, and this change is represented by a path on the liquidus projection, and occurs in a direction representing a lowering of temperature on the liquidus surface. This path is formed by the extremities of the tie-lines representing the liquid compositions in equilibrium with the primary phase at successively lower temperatures.

When the primary phase is of fixed composition, e.g. a pure metal, or an intermediate phase showing no solid solubility, the path is simply the extension of the straight line joining the composition of the primary phase and the alloy composition. At any temperature during the primary solidification stage, if the liquid composition is known, the amounts of liquid and solid can be calculated by applying the lever rule to the tie-line. Thus the tie-line AXZ in Fig. 4.4, represents the equilibrium at the end of the primary solidification stage when the liquid composition has just reached a point on the eutectic curve dg; from this tie-line, the amount of primary A is represented by the ratio XZ/AZ.

In systems where the primary phase is a solid solution, its composition changes during solidification and the accurate tracing of the paths along which the liquid and solid compositions change requires experimental data regarding tie-lines; these data are not usually available. However, when the extent of the solid solubility is small, the procedure for a phase showing no solid solubility may be used as an approximation, i.e. taking the extension of the straight line joining the composition of the pure metal or intermediate phase. With extensive solid solubility, however, the adoption of this procedure may be quite misleading. It is then preferable to attempt to estimate the change in liquid and solid compositions with reference to the solidus projection, which will at least help in locating the likely compositions of liquid and solid co-existing at the beginning of a three-phase reaction immediately following the primary solidification.

It should be noted that where solid solubility exists it is not possible to calculate exactly the relative amounts of liquid and solid co-existing during primary solidification unless tie-line data are available.

Curves on a liquidus projection

A curve such as, for example, a eutectic valley, represents the path along which the liquid composition changes while undergoing a three-phase solidification reaction; the liquid co-exists with two solids in a eutectic or peritectic reaction, and with a solid and another liquid in a monotectic reaction.

The three-phase reaction commences when the path representing the changing liquid composition during the primary stage intersects a curve on the liquidus projection. The three-phase reaction then proceeds as the liquid composition changes along the curve; commonly, arrows are shown on the curves to indicate the direction of decreasing temperature. With regard to the solid phases separating during the three-phase reaction, their compositions will be fixed in the case of pure metals or other components showing no solid solubility. When the phases separating are solid solutions, their compositions change along the relevant solid solubility curves; detailed experimental data would be required to locate the tie-triangles which would be necessary for a calculation of the amounts of liquid and solid existing at the various temperatures, using the centre-of-gravity rule.

Reference to the limits of the solid state regions (as shown on a solidus projection, or isothermal section), will show if the solidification of a given alloy is completed by the three-phase reaction to give a two-phase alloy (e.g. as in alloy X of Fig. 4.3). Alternatively, the liquid may undergo an invariant reaction as a sequel to the three-phase reaction, (e.g. as in alloy Y of Fig. 4.3 which consists of $\alpha + \beta + \gamma$ when solid, and which therefore must complete its solidification by the ternary eutectic at *g*).

Points on a liquidus projection

Points formed by the intersection of curves on a liquidus projection represent the composition of liquid involved in an invariant reaction. Such a point is formed by the intersection of three liquidus curves, and when the liquid composition reaches it, the invariant reaction occurs. Assuming equilibrium conditions, the temperature remains constant until the reaction is completed, and during the reaction the compositions of all four phases involved remain unchanged.

The nature of the particular reaction represented by a point may be deduced by reference to the arrows shown on the intersecting liquidus curves; the three examples discussed in the preceding chapters are as follows:

(i) Ternary eutectic $L \rightarrow \alpha + \beta + \gamma$

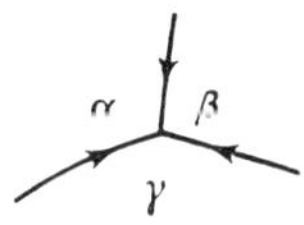

(ii) Ternary peritectic $L + \alpha \rightarrow \beta + \gamma$

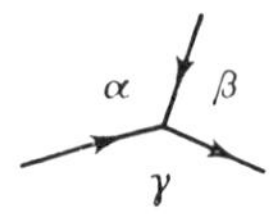

(iii) Ternary peritectic $L + \alpha + \beta \rightarrow \gamma$

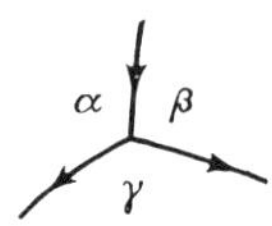

With peritectic types of reaction, reference to the solid state constitution will reveal whether a given alloy composition lies in the three-phase $\alpha + \beta + \gamma$ region, in which case solidification is completed by the invariant reaction. If the composition of an alloy that undergoes the invariant reaction lies outside the $\alpha + \beta + \gamma$ region, liquid will remain at the completion of the invariant reaction (see for example region *ngo* of Fig. 4.11). This liquid will then undergo a three-phase reaction along a curve running from the invariant point to lower temperatures, and assuming that the system does not contain any other invariant reactions the alloy will solidify as a two-phase mixture, or possibly, for the $L + \alpha + \beta \longrightarrow \gamma$ reaction, a single phase γ structure may result (e.g. as in region *ogp* of Fig. 4.15).

If the system contains another invariant reaction, involving an additional solid phase, say δ, the liquid composition may reach the second invariant point before solidification is complete; reference to the solid state constitution will show whether this will be so, e.g. see Fig. 5.9.

It should be noted that the reaction corresponding to example (i) above is a ternary eutectic, irrespective of whether α, β and γ are primary solid solutions or intermediate phases. However, if a system contains allotropic transformations or solid state transformations involving intermediate phases, then additional information is needed to identify with certainty the invariant reactions represented by the liquidus arrangements in examples (ii) and (iii). Other types of reaction that may occur may be found in Reference 7, Chapter 18.

Non-equilibrium cooling

When the cooling rate is too rapid to allow equilibrium to be attained, 'coring' effects similar in principle to those occurring in binary systems will often arise.[7] These can, for example, produce a three-phase structure, where a two-phase structure should exist under equilibrium: thus, alloy X of Fig. 4.3 might contain some γ phase, due to the liquid composition reaching point g under non-equilibrium cooling conditions. The effects of non-equilibrium cooling may be particularly marked in systems containing peritectic reactions, due to the reactions not proceeding to completion. For example, in the system represented in Fig. 5.9, alloys lying in the $\alpha + \delta + \gamma$ region when in equilibrium, might contain some β phase after rapid cooling, due to the invariant peritectic reaction at h failing to go to completion, and the composition of the remaining liquid eventually reaching point g to undergo the ternary eutectic reaction. The aluminium–copper–magnesium system depicted in Figs. 5.18*a* and *b* illustrates this type of situation for the non-equilibrium cooling of certain alloys. For example, an alloy containing 15% Cu and 17% Mg should complete its solidification by the invariant peritectic reaction at point f to give Al + T + S; however, with rapid cooling the liquid composition can reach the ternary eutectic point h representing the reaction $L \longrightarrow \text{Al} + \beta + \text{T}$ so that, in addition to Al, T, and S, the solid alloy will contain some β phase.

7
Selected Case Studies of Ternary Systems

C–Nb–Ta

This system provides an example of complete mutual solid solubility between two refractory metal monocarbides, of interest for their high melting points and hardness. The space model depicted in Fig. 7.1[13] shows the Nb–C and Ta–C systems as being of the same type, containing

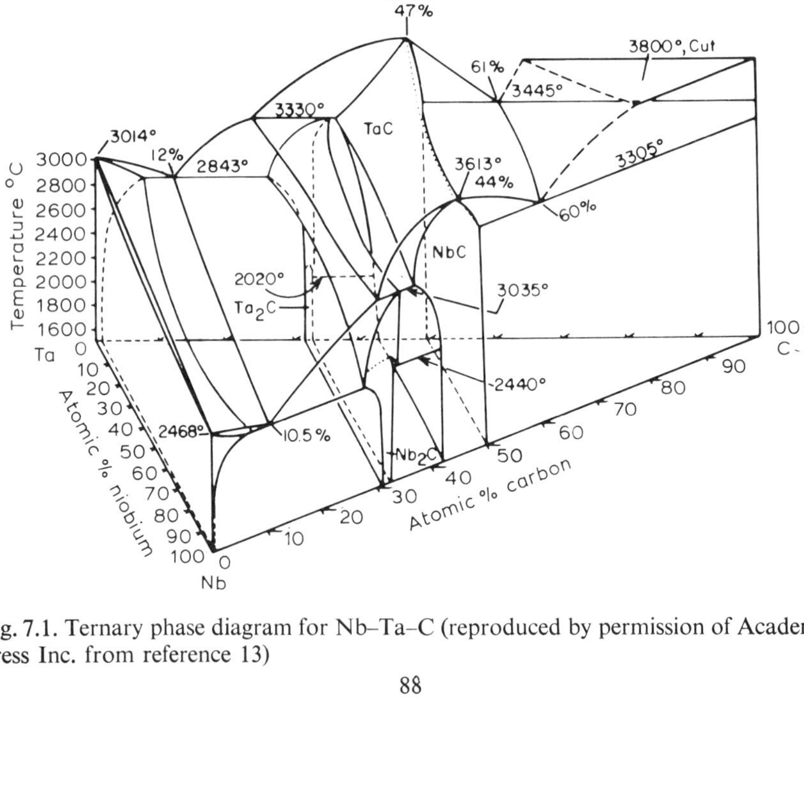

Fig. 7.1. Ternary phase diagram for Nb–Ta–C (reproduced by permission of Academic Press Inc. from reference 13)

carbide phases MC and M_2C (the latter has two structural modifications). Regions of complete solid solubility link Nb and Ta, Nb_2C and Ta_2C; and NbC and TaC, respectively. The following three-phase reactions occur: $L \leftrightharpoons$ Nb–Ta solid solution + M_2C; $L +$ MC $\leftrightharpoons M_2C$; $L \leftrightharpoons$ MC + C. The diagram is probably an oversimplification since the literature shows differences in detail of the carbide phase relationships.

Problem

Draw a schematic vertical section from the mid-point of the Nb–Ta system to the C corner of the system and label the phase regions. By reference to this section describe the changes that occur during the melting of an alloy containing 55 at% Nb, 10 at% Ta and 35 at% C.

As–Ga–Zn

Ternary phase diagrams for systems of Ga–As + an element such as Zn, Sn, Cu, Ag and Au are important in relation to the incorporation of a dopant material into GaAs to produce p–n junctions. Consider the Ga–As–Zn system[14, 15] (Fig. 7.2). Elemental Zn, dilute solutions of Zn in Ga or various combinations of Zn and As can be used as source materials in sealed capsules, for the diffusion of Zn into the compound. However, alternative source materials have been investigated in the light of phase diagram data.[16,17] These include compositions lying in a three-phase region (e.g. 5Ga–50As–45Zn (at%) in the GaAs + Zn_3As_2 + $ZnAs_2$ solid state region at temperatures below ~744°C); within such regions the partial pressures of the components do not vary with composition so that the surface composition and diffusivity depend only on temperature. Such diffusion sources have been found advantageous with respect to obtaining reproducible steep diffusion profiles and planar junction interfaces; also it is not necessary to maintain stringent control of the system composition and capsule volume.

Problems

1. Consider a diffusion source of composition (at%) 50Ga–25As–25Zn placed with a sample of GaAs in an evacuated and sealed silica capsule, and heated to 1000°C. Deduce the composition of the phase in equilibrium with GaAs at this temperature.

2. (a) Draw an isothermal section for the system at 700°C. (b) By reference to this section calculate the proportions by weight of the phases present in an alloy containing 5Ga–50As–45Zn (at%) equilibrated at 700°C.

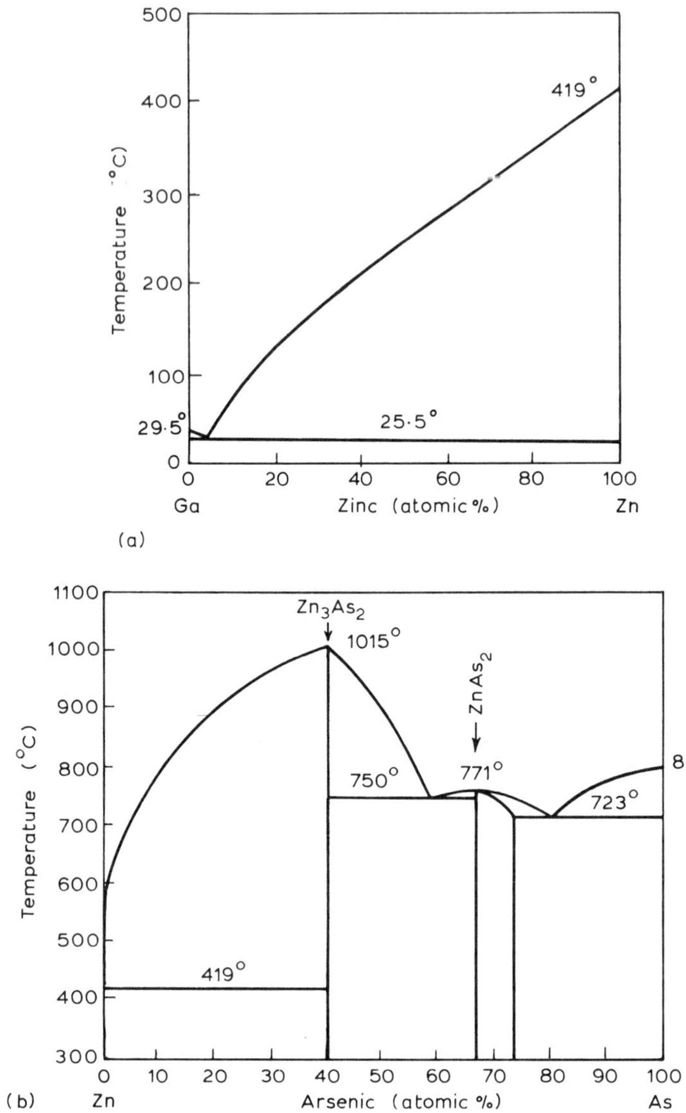

Fig. 7.2 (*a*) Ga–Zn binary system (*b*) Zn–As binary system (*c*) Ga–As binary system (*d*) Liquidus surface of As–Ga–Zn system[14, 17] representing the condensed phases in equilibrium with vapour (under closed system conditions). Three congruently melting binary compounds exist: GaAs, Zn_3As_2 and $ZnAs_2$, with associated eutectic reactions: $L \rightleftharpoons$ Ga + GaAs (29·5°C); $L \rightleftharpoons$ GaAs + As (810°C); $L \rightleftharpoons$ Zn + Zn_3As_2 (419°C); $L \rightleftharpoons Zn_3As_2 + ZnAs_2$ (750°C); $L \rightleftharpoons ZnAs_2$ + As (723°C). The Ga–Zn system contains a eutectic reaction: $L \rightleftharpoons$ Ga + Zn (25·5°C).

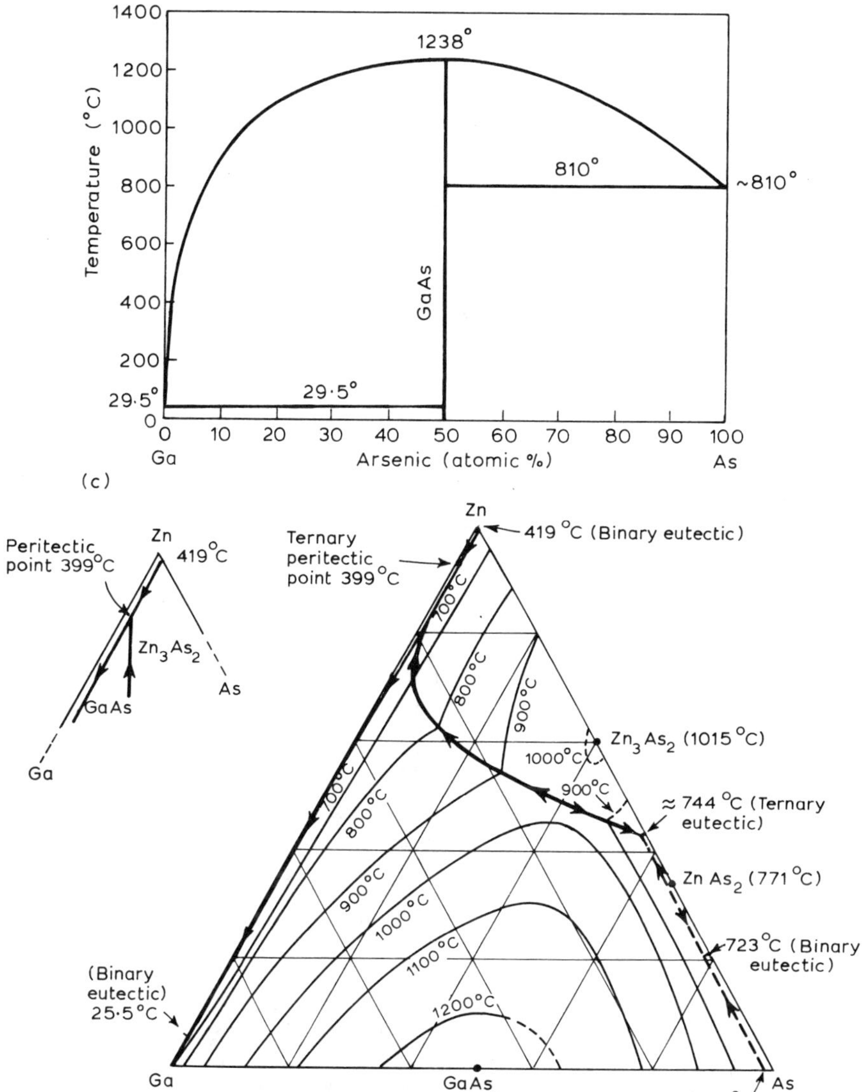

Fig. 7.2 (*cont.*)

There are four ternary invariant reactions in each of which the liquid compositions lie very close to binary compositions: $L \rightleftharpoons$ Ga + Zn + GaAs (~20°C); $L + Zn_3As_2 \rightleftharpoons$ GaAs + Zn (399°C); $L \rightleftharpoons$ GaAs + Zn_3As_2 + $ZnAs_2$ (~ 744°C); $L \rightleftharpoons$ GaAs + $ZnAs_2$ + As (~720°C). The mutual solid solubilities are very small for all the solid phases in the system

(*d* reproduced by permission of Pergamon Press Ltd from reference 14)

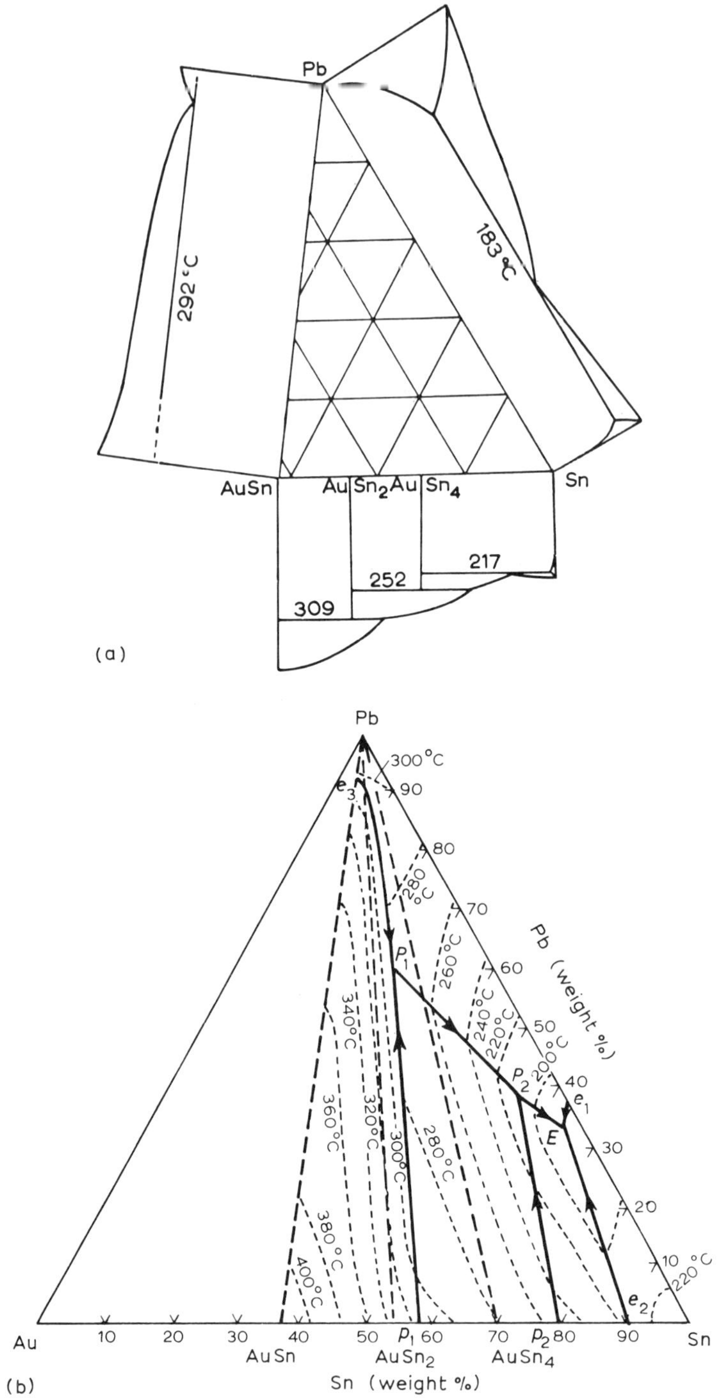
Pb
292 °C
183 °C
AuSn
AuSn2
AuSn4
Sn
217
252
309
(a)
Pb
300 °C
e3
90
80
280 °C
70
260 °C
60
P1
340 °C
240 °C
220 °C
50
200 °C
Pb (weight %)
P2
40
e1
E
360 °C
320 °C
300 °C
280 °C
30
380 °C
20
400 °C
10
220 °C
e2
Au
10
20
30
40
50
p1
60
70
p2
80
90
Sn
AuSn
AuSn2
AuSn4
Sn (weight %)
(b)

3. From the nature of the reactions at 744°C and 720°C what can be deduced about the equilibria along the sections GaAs–Zn_3As_2 and GaAs–$ZnAs_2$?

Au–Pb–Sn

The Au–Pb–Sn system[18] (Fig. 7.3), particularly the region between Pb, Sn and the compound $AuSn_2$, is relevant to the use of Pb–Sn solders for joining Au-plated surfaces in the electronics and telecommunications industries. The liquidus has a relatively shallow slope near the Pb–Sn binary system so that substantial solution of Au occurs in molten solder at normal soldering temperatures. The formation of platelets of the brittle compound $AuSn_4$ occurs on solidification, involving the invariant peritectic (P_2) (which does not proceed to completion at the cooling rates encountered in soldering) and the eutectic solidification (E). One procedure for minimizing embrittlement is to keep the thickness of the Au coating below 5 μm.

Problems

1. Describe the solidification under equilibrium conditions of an alloy containing 10 wt% Au, 40 wt% Pb, and calculate the proportions of the phases present at 170°C, assuming negligible solid solubility.

2. What percentage of Au can dissolve in a 40% Pb–60% Sn solder at 280°C?

3. Compare the equilibrium solidification behaviour of a 40% Pb, 60% Sn solder used on a gold-plated surface at 220°C with that of a 50% Pb, 50% Sn solder used at 240°C; include reference to the proportions of liquid that undergo the peritectic reaction at P_2 and the proportions of $AuSn_4$ that will form in the two alloys.

Fig. 7.3 (*a*) The constituent binary phase diagrams of the AuSn–Pb–Sn ternary system[18]
(*b*) Liquidus surface of the AuSn–Pb–Sn system involving the solid solutions based on Pb and Sn and the intermetallic compounds AuSn, $AuSn_2$ and $AuSn_4$[18]. The Pb–Sn system contains a eutectic (e_1) (183°C). AuSn is a congruently melting compound (419°C), and the binary system between AuSn and Sn contains the following reactions: $L + AuSn \leftrightharpoons AuSn_2$ (p_1) (309°C); $L + AuSn_2 \leftrightharpoons AuSn_4$ (p_2) (252°C); $L \leftrightharpoons AuSn_4 + Sn$ (e_2) (217°C). The section AuSn–Pb is quasi-binary containing the eutectic $L \leftrightharpoons AuSn + Pb$ (e_3) (292°C).
Ternary invariant reactions occur as follows: $L + AuSn \leftrightharpoons Pb + AuSn_2$ (P_1) (275°C); $L + AuSn_2 \leftrightharpoons Pb + AuSn_4$ (P_2) (208°C); $L \leftrightharpoons Pb + Sn + AuSn_4$ (E) (177°C). Solid solubilities are neglected here. Dashed lines (— — — —) represent the sections between Pb and the various Au–Sn compounds
(*a* and *b* reproduced by permission of Elsevier Sequoia from reference 18)

Al–Cr–Ni

Nickel-based superalloys are of complex constitution, and contain a range of alloy elements, typically including Cr, Al, Ti, Co, Mo and C. These alloys consist essentially of a nickel-based solid solution (γ), an intermetallic compound based on Ni_3Al (γ'), and carbide phases. The intermetallic sigma (σ) phase may also be present, although alloys are designed to avoid its formation. The ternary system Ni–Al–Cr[19] (Figs. 7.4*a–g*) forms a useful basis for understanding the relationship between the γ and γ' phases, which are of major importance in the control of mechanical properties: it is also relevant to the role of the β-phase as a constituent of certain surface coatings on superalloy components.

Superalloys typically form primary γ on solidification and commonly contain some $\gamma + \gamma'$ eutectic: they are heat-treated to give precipitation of γ' from supersaturated γ. The isothermal sections at 1000°C and 750°C (Figs. 7.4*e* and *f*), together with the vertical section at 75 at% Ni (Fig. 7.4*g*) illustrate the extensive composition range of γ, and the decrease in solubility of Al in γ with decreasing temperature.[19] Consider, for example, an alloy containing 75 at% Ni, 10 at% Al, 15 at% Cr, solution treatment at, say, 1000°C, followed by rapid cooling, produces supersaturated γ, and ageing at, say, 750°C, produces a fine dispersion of γ' precipitate.

The role of Ti can be taken into account by reference to a section through the quaternary Ni–Cr–Al–Ti system at 75 at% Ni and at 750°C (Fig. 7.4*h*).[20] (A quaternary isothermal, isobaric section can be represented by an equilateral tetrahedron, and a section at a constant proportion of one component can be represented by an equilateral triangle.) In Fig. 7.4*h* the γ/γ' tie lines lie close to the plane of the section, and the composition point X is close to that of Nimonic 80A (i.e. ~72 Ni, 21 Cr, 2.7 Al, 2.9 Ti (atomic %) + small amounts of other elements). The γ field extends to higher Al and Ti contents with increasing temperature, and at 1000°C Nimonic 80A is well within the γ region for solution treatment. An important trend in wrought and cast superalloy development (particularly in cast alloys) has been to increase the γ' content by increasing the Al + Ti content.

To take account of the influence of other elements on the constitution of superalloys, various ternary and quaternary systems can be consulted. For example, the influence of refractory metals on the γ/γ' equilibrium may be illustrated by systems such as Ni–Al–Mo and Ni–Al–Nb. Sigma-phase formation can be considered in terms of systems such as Ni–Cr–Co and Ni–Cr–Co–Mo, and carbide formation in terms of systems such as Ni–Cr–Mo–C. While such ternary and quaternary phase diagrams provide valuable insights, they do not represent precisely the combined effects of

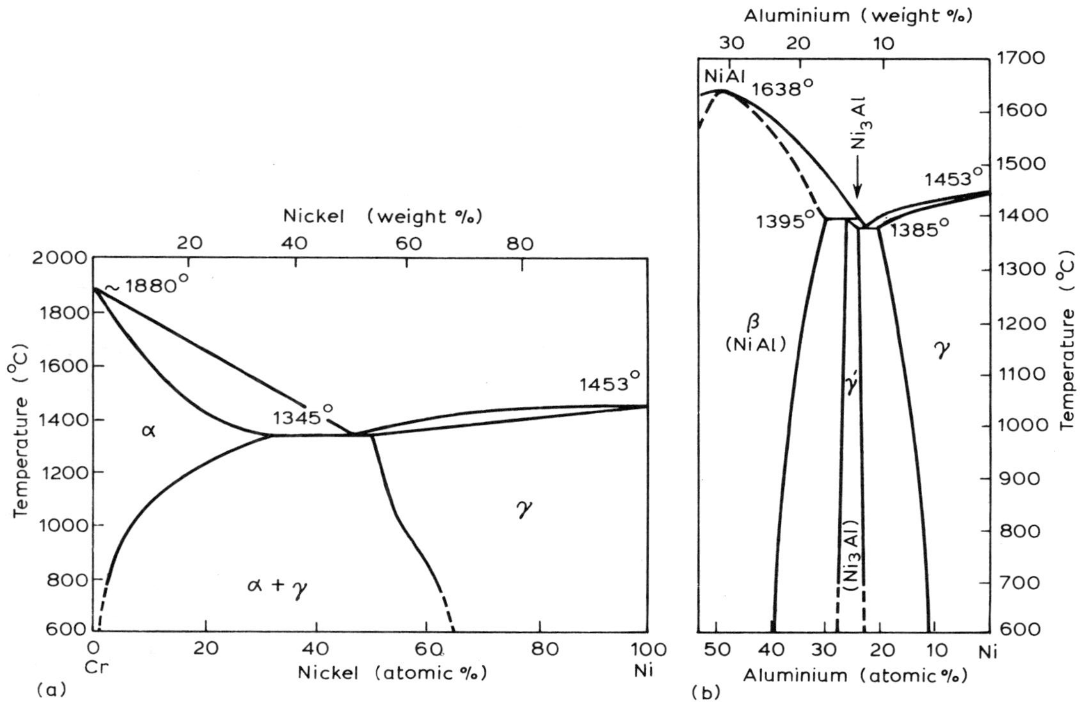

Fig. 7.4 (*a*) Cr–Ni binary system. Note that the α-range data differ from the data used in (*c*) (e.g. showing a greater solubility at the eutectic temperature) being taken from a more recent source
(*b*) Part of Ni–Al binary system

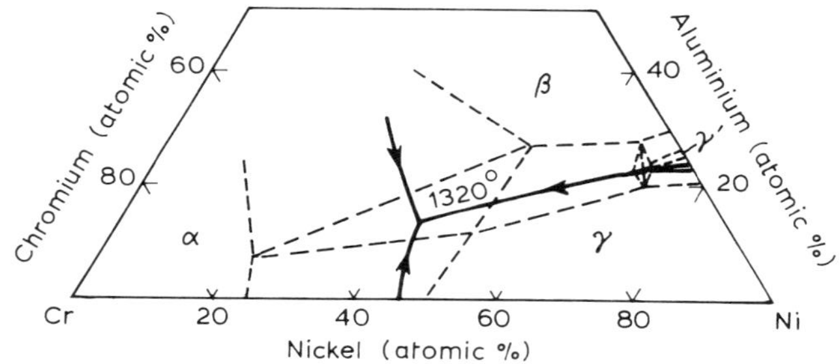

(c)

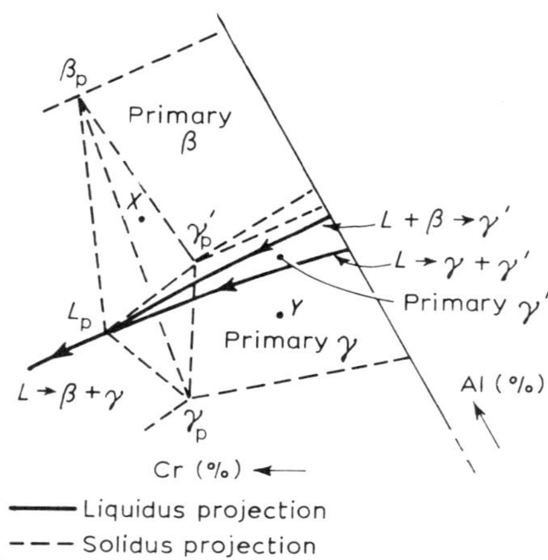

(d)

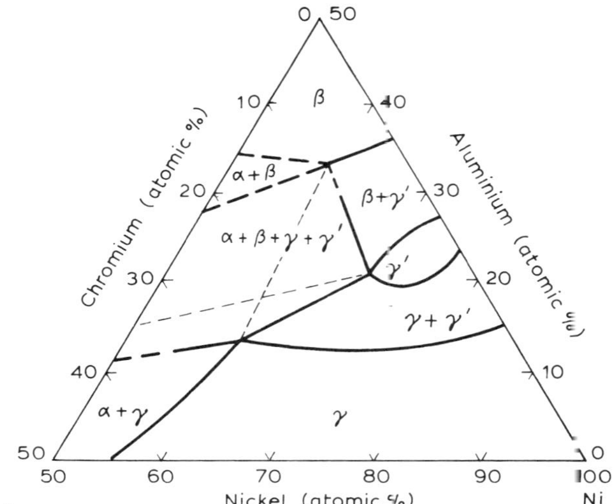

(e)

Fig. 7.4 (*cont.*)

(*c*) Tentative diagram of fields of primary crystallization for the portion of the Ni–Cr–Al system involving γ (Ni-based solid solution), γ′ (Ni_3Al), β (NiAl) and α (Cr-based solid solution). The invariant liquid/solid reactions are: *Binary*: $L + \beta \rightleftharpoons \gamma'$ (1395°C); $L \rightleftharpoons \gamma + \gamma'$ (1385°C); $L \rightleftharpoons \alpha + \gamma$ (1345°C). *Ternary*: $L + \gamma' \rightleftharpoons \beta + \gamma$ (1340°C); $L \rightleftharpoons \alpha + \beta + \gamma$ (1320°C). Also a solid state invariant reaction, $\beta + \gamma \rightleftharpoons \alpha + \gamma'$ occurs at ~ 1000°C (Figs 7.4*e* and A9). The dashed lines represent the solid state constitution including the invariant reaction planes at 1340°C and 1320°C

(*d*) Enlarged view of part of the Ni–Cr–Al system (Fig. 7.4*c*) showing the $L + \gamma' \rightleftharpoons \beta + \gamma$ reaction

(*e*) Isothermal section at 1000°C

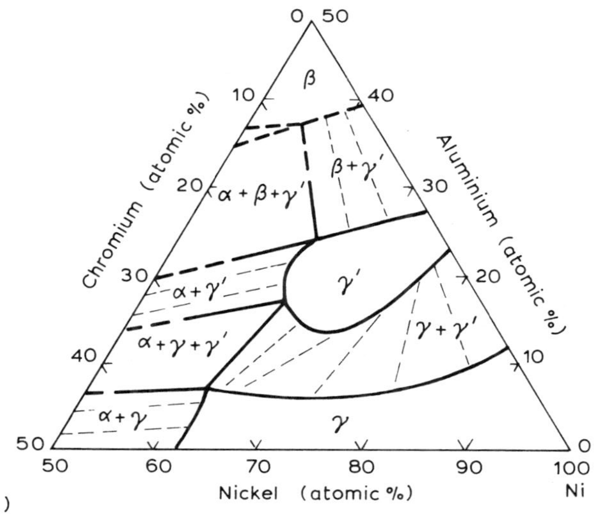

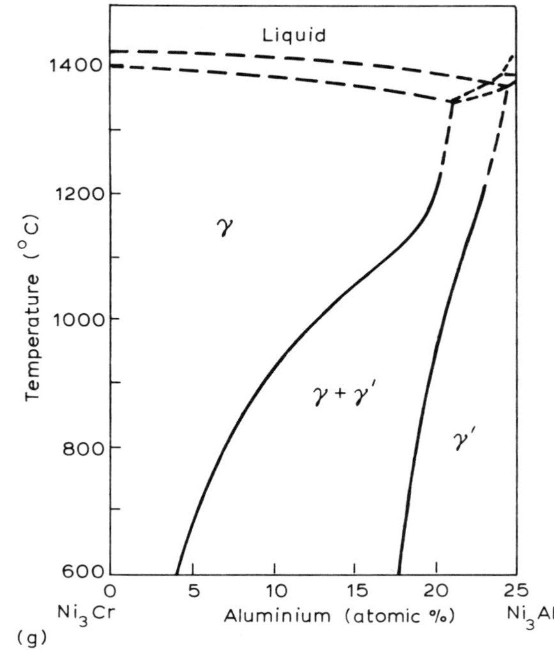

Fig. 7.4 (*cont.*)
(*f*) Isothermal section at 750°C
(*g*) Vertical section at 75 at % Ni

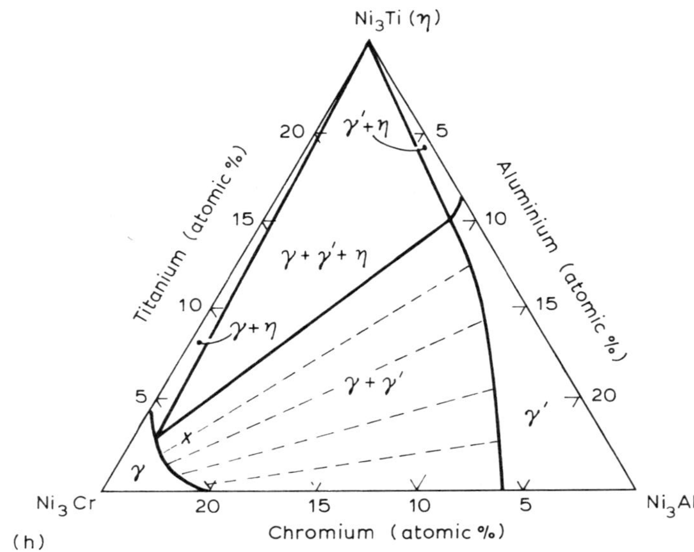

Fig. 7.4 (*cont.*)

(*h*) The Ni_3Cr–Ni_3Ti–Ni_3Al section of the Ni–Cr–Ti–Al system; 750°C isothermal. X corresponds approximately to the composition of Nimonic 80A alloy

the wide range of elements commercially involved. The PHACOMP procedure for controlling sigma formation and the use of 'polar' phase diagrams are complementary approaches to the understanding of industrial superalloys.[21]

Problems

1. By reference to Fig. 7.4*d* describe the equilibrium solidification sequences of (a) an alloy that initially deposits primary β, and which consists of $\beta + \gamma + \gamma'$ when solidification is completed, and (b) an alloy containing 75 at% Ni, 22 at% Al and 3 at% Cr.

2. Draw the three phase regions lying just above and just below the invariant reaction plane at 1000°C to illustrate the transition from the $\beta + \gamma$ to the $\alpha + \gamma'$ equilibrium.

3. Discuss the effect on the solid state equilibria of increasing the Cr content from 0–25 at% in series of alloys containing (a) a constant Al content of 15 at%, and (b) a constant Ni content of 75 at%.

C–Cr–Fe, Cr–Fe–Ni and C–Cr–Fe–Ni

The iron-rich corner of the C–Cr–Fe–Ni system forms the basis of the wide range of alloys designated 'stainless steels'. The ternary systems C–Cr–Fe and Cr–Fe–Ni provide a basis for understanding the constitution of these materials, which can be extended by appropriate sections through the quaternary system. Solid state transformations involving the allotropic forms of iron, ferrite (α or δ), bcc, and austenite (γ), fcc, are of major importance. The transition between α-stabilizing (Fe–Cr) and γ-stabilizing (Fe–C and Fe–Ni) phase relationships gives rise to complex regions in the ternary systems.[5, 22]

The binary systems Fe–Cr, C–Cr, Fe–Ni and Cr–Ni are represented in Figs 7.5*a–c* and 7.4*a*. In the Fe–Cr system all alloys solidify as bcc α-phase; the system shows a γ-loop region and beyond 13% Cr γ does not form. Below ~820°C sigma-phase (σ) forms, centred around the approximate equiatomic composition and entering into equilibrium with α over a wide composition range with an associated embrittling effect. At ~440°C σ decomposes eutectoidally into iron-rich and chromium-rich bcc solid solutions (the chromium-rich solution may be designated α'). The Fe–Ni system contains a peritectic reaction $L + \delta \leftrightharpoons \gamma$. The $\gamma \leftrightharpoons \alpha$ transformation temperature is progressively decreased by Ni additions. The Cr–Ni system shows extensive solid solubility of Cr in Ni (γ-phase) and of Ni in Cr (α-phase) and there is a eutectic reaction $L \leftrightharpoons \alpha + \gamma$.

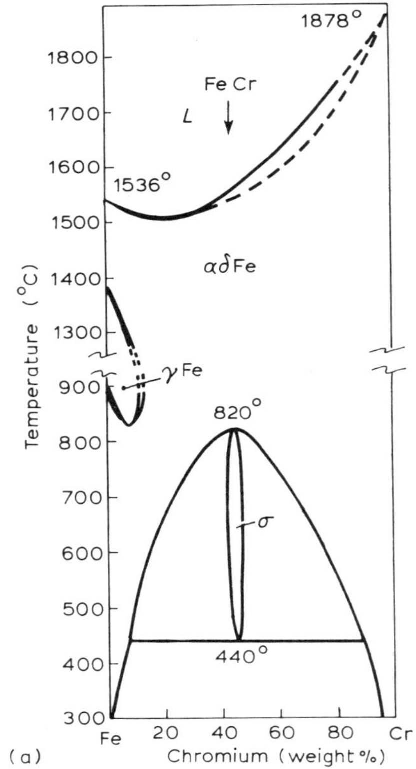

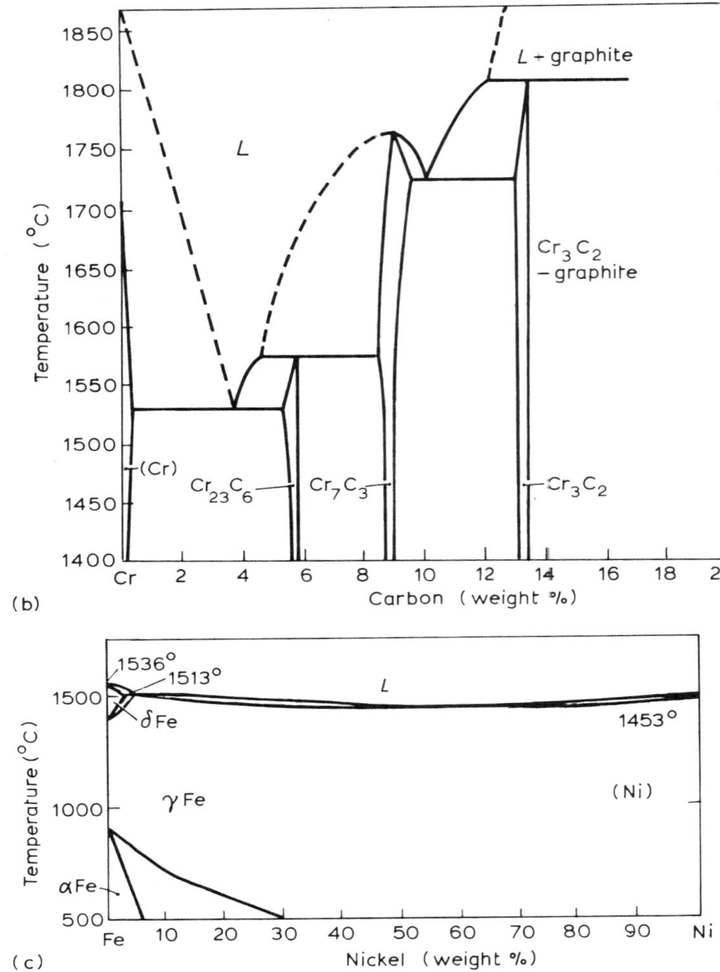

Fig. 7.5 (*a*) Cr–Fe binary system
(*b*) Part of C–Cr binary system
(*c*) Fe–Ni binary system (certain solid state transformations that occur at less than approximately 560°C are not shown here)

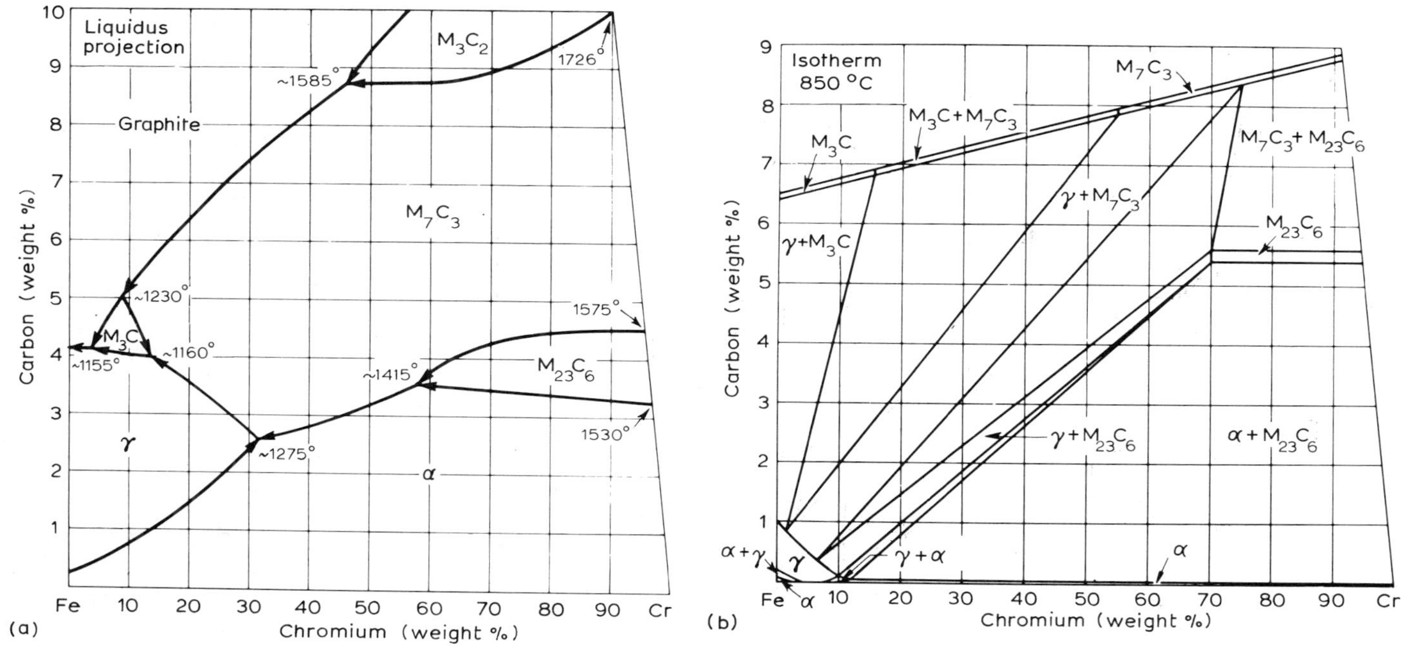

Fig. 7.6 C–Cr–Fe system[23, 24]

(*a*) Liquidus projection
(*b*) Isothermal section at 850°C

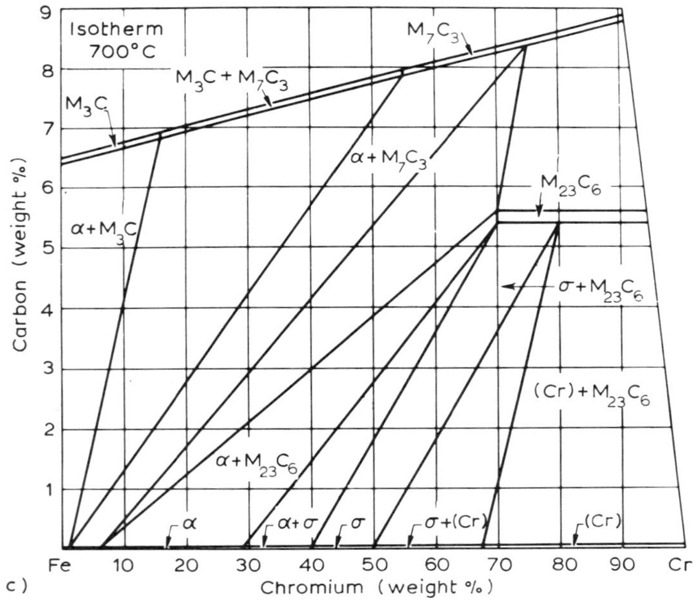

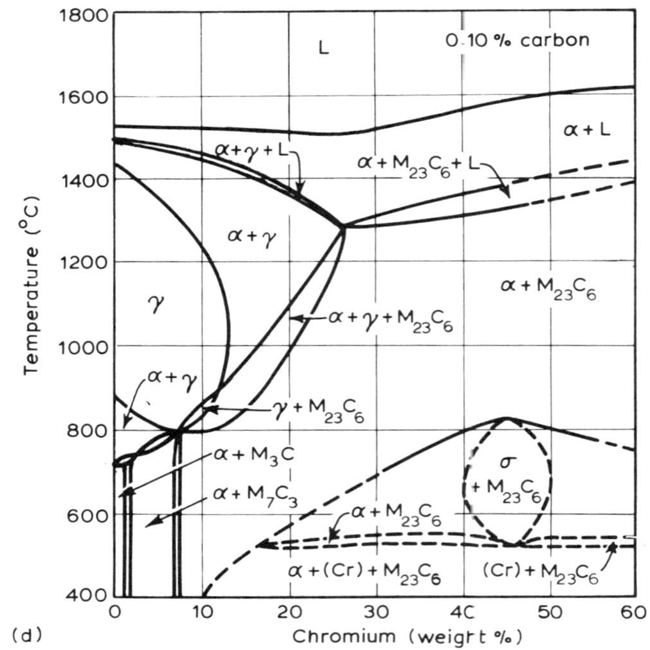

Fig. 7.6 (*cont.*)

(*c*) Isothermal section at 700°C
(*d*) Vertical section at 0.10% C
(*a* to *d* reproduced by permission of the American Society for Metals from reference 23)

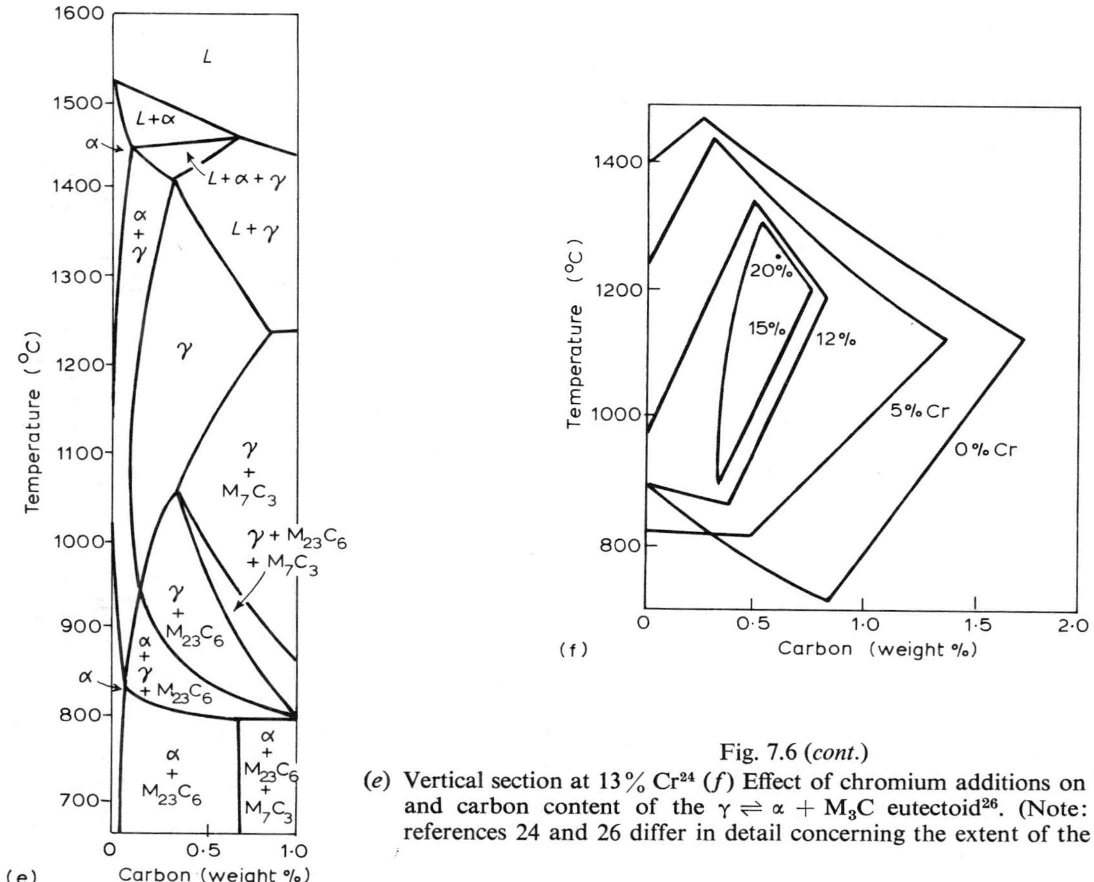

Fig. 7.6 (*cont.*)

(*e*) Vertical section at 13% Cr[24] (*f*) Effect of chromium additions on the temperature and carbon content of the $\gamma \rightleftharpoons \alpha + M_3C$ eutectoid[26]. (Note: the data from references 24 and 26 differ in detail concerning the extent of the γ region.)

C–Cr–Fe

The liquidus projection[23] (Fig. 7.6*a*) shows the following phases: α, γ, graphite, M_3C (based on the Fe–C system), and three carbides ($M_{23}C_6$, M_7C_3 and M_3C_2) based on the Cr–C system and showing extensive solubility for Fe. The solid state constitution is illustrated by isothermal sections at 850°C (Fig. 7.6*b* shows the equilibria between the γ-phase and the carbides) and at 700°C (Fig. 7.6*c* shows the equilibria between the α-phase and the carbides); equilibria do not occur between M_3C_2 and either γ or α, because of the very high solubility of Fe in M_7C_3. Vertical sections are shown at constant C (0.1%) with variable Cr, and at constant Cr (13%) with variable C[24, 25] (Figs. 7.6*d* and *e*). The solid state constitution includes two invariant reactions:

$$\gamma + M_{23}C_6 \leftrightarrows \alpha + M_7C_3$$
$$\gamma + M_7C_3 \leftrightarrows M_3C + \alpha$$

With the addition of C to Fe–Cr alloys the γ-loop is moved to higher Cr levels and the $\alpha + \gamma$ field is widened. The maximum displacement of the phase boundary for fully austenitic alloys is reached at 0.6% C and 18% Cr. Considering the addition of Cr to Fe–C alloys, the γ field is constricted, the eutectoid temperature ($\gamma \leftrightarrows \alpha + M_3C$) being raised and the C content of the eutectoid being lowered (Fig. 7.6*f*).[26]

Problems

1. State the nature of each of the ternary invariant reactions shown in Fig. 7.6*a*.

2. Describe the equilibrium solidification sequence to just below the solidus of an Fe–13 wt% Cr–0.1 wt% C alloy making reference to Figs 7.6*a* and *d*.

3. Samples of an Fe–13 wt% Cr–0.5 wt% C alloy are heated for one hour at temperatures of 1200°C and 850°C respectively. The samples are then water quenched and tempered for one hour at 700°C. Describe the structural changes that would occur during these treatments, making reference to Figs 7.6*b* and *e*.

Cr–Fe–Ni

This system has been extensively investigated and recently the data have been critically evaluated.[27] The liquidus and solidus projections for the ternary Cr–Fe–Ni system (Figs 7.7*a* and *b*) show the transition between the peritectic reaction in the Fe–Ni system and the eutectic in the Cr–Ni system; and there is a temperature minimum not far from the Cr–Ni binary.

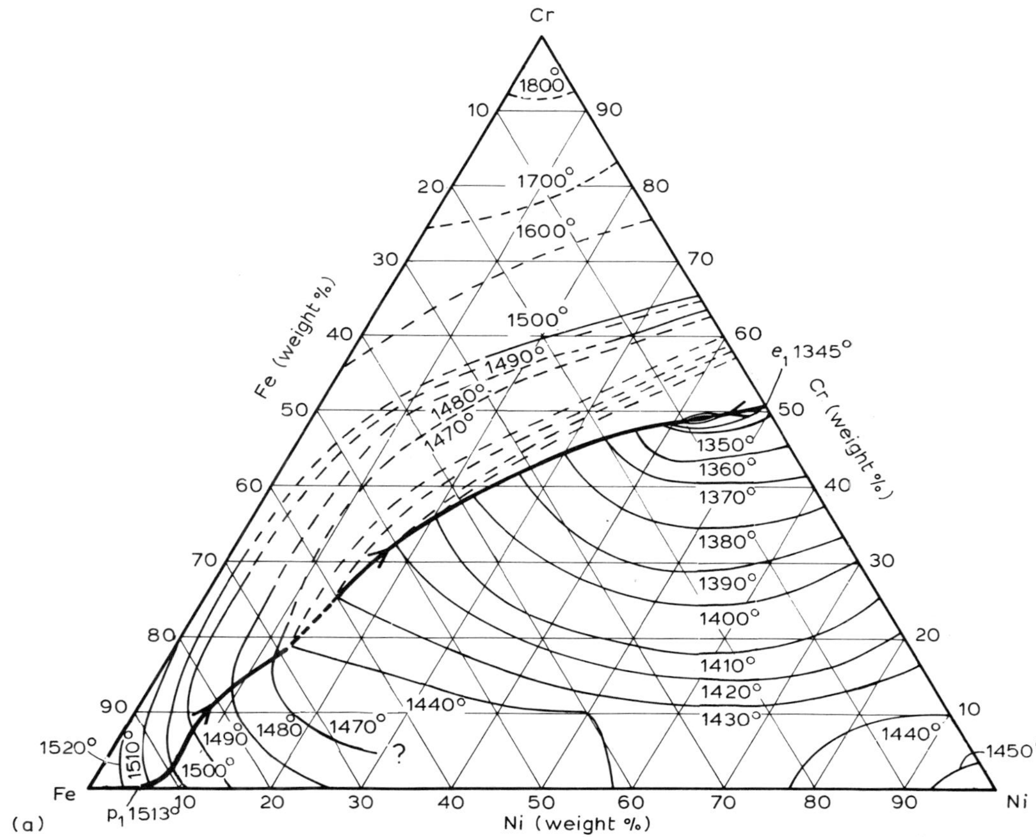

Fig. 7.7. Cr–Fe–Ni system[27]

(*a*) Liquidus projection

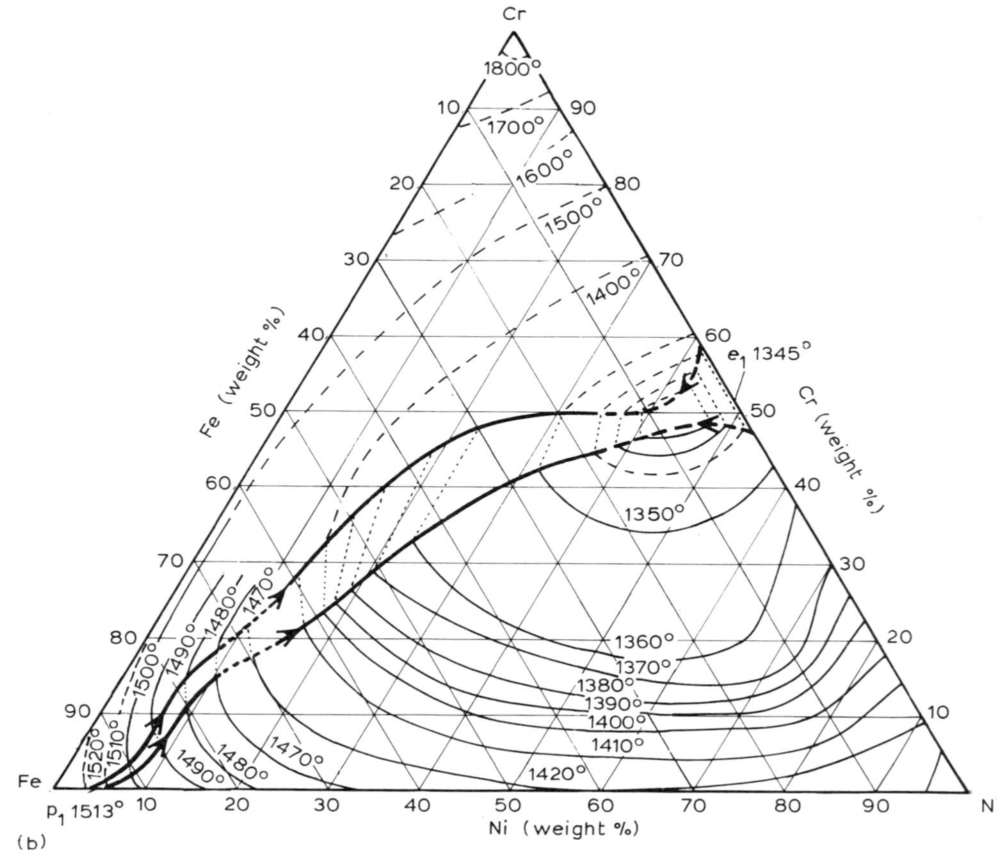

Fig. 7.7 (*cont.*)

(*b*) Solidus projection

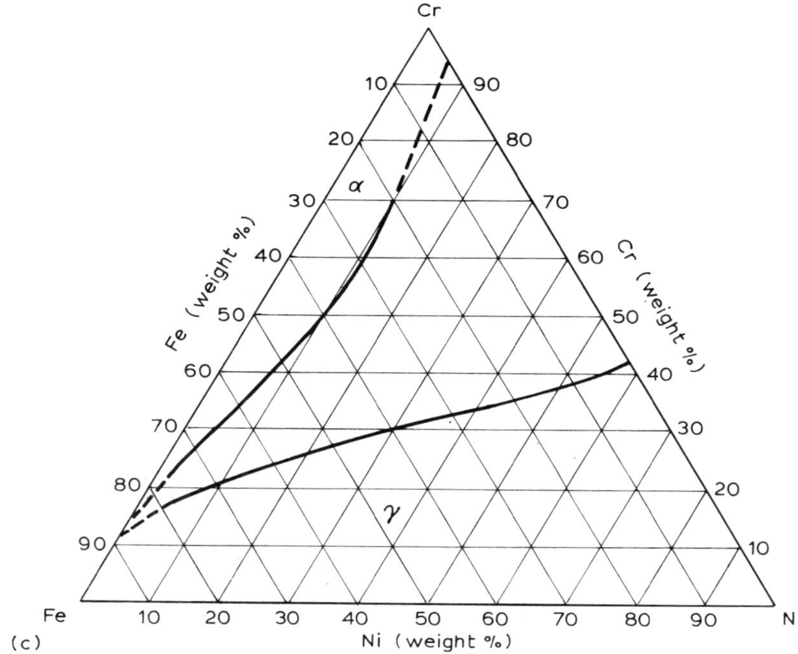

Fig. 7.7 (*cont.*)

(*c*) Isothermal section at 1000°C

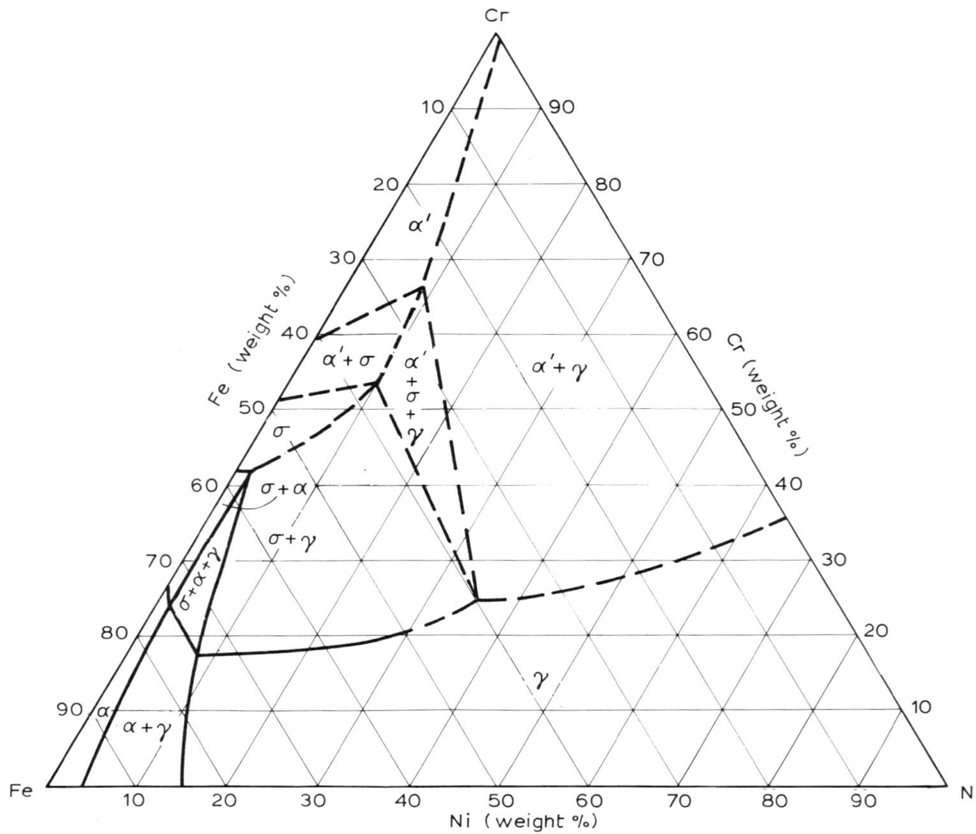

Fig. 7.7 (*cont.*)

(*d*) Isothermal section at 650°C

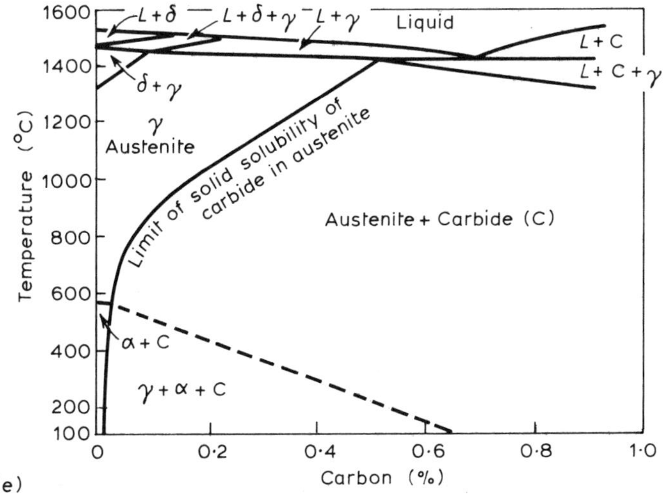

(e)

Fig. 7.7 (*cont.*)

(*e*) Vertical section showing effect of carbon additions to Fe–18Cr–8Ni steels (C = carbide)[28]
(*a* to *d* reproduced by permission of the Metals Society from reference 27)

The isothermal section at 1000°C (Fig. 7.7*c*) shows the extensive single phase α and γ regions, together with the two-phase region linking the α + γ region bounding the γ-loop in the Fe–Cr system with the α + γ region formed by eutectic solidification in the Cr–Ni system. This section forms part of the basis of the Schaeffler diagram for representing the constitution of stainless steels, including the effect of additions of various alloy elements, and used, for example in relation to welding. At temperatures below about 900°C (e.g. at 650°C, Fig. 7.7*d*) the α + γ region extends into the ternary system from the Fe–Ni and Cr–Ni systems; in the Fe–Ni system the α + γ/γ boundary moves to higher nickel contents with decrease in temperature. The upper temperature limit for σ formation is ~950°C and the equilibria between σ, α (α′) and γ at 650°C are shown in Fig. 7.7*d*.

C–Cr–Fe–Ni

The effect of C on the constitution of an Fe–18Cr–8Ni steel is illustrated by the vertical section in Fig. 7.7*e*.[28] Rapid quenching from the γ region of the Fe–18Cr–8Ni material with typical carbon levels (e.g. ~0.05–0.1 wt %) gives a retained austenite structure. Heating within the γ + carbide region can produce carbide precipitation; the preferential nucleation of carbide at the austenite grain boundaries leads to chromium depletion of the adjoining regions with adverse effects on corrosion resistance (e.g. as in 'weld decay').

Problems

1. Draw an isothermal section for the Cr–Fe–Ni system at 1400°C, based on the data in Figs 7.7*a* and *b* and making necessary assumptions where relevant data are lacking.

2. Describe the equilibrium solidification sequence of an alloy containing 30 wt% Cr and 15 wt% Ni, including a calculation of the proportions of the phases present at 1400°C.

3. Describe the changes that occur in an Fe–18 wt% Cr–12 wt% Ni alloy (i.e. an 'austenitic' stainless steel) during the solidification and subsequent cooling to 650°C assuming that equilibrium is attained. For this same alloy state what structure would result from rapid cooling from 1000°C. (Note: the M_s is below room temperature).

4. Consider a series of Cr–Fe–Ni alloys containing 5 wt% Ni and Cr contents ranging from ~20–70 wt%, annealed to equilibrium at 650°C. Describe how the proportion of σ-phase varies with Cr content, making reference to Fig. 7.7*d*.

Fe–O–S

The system (Fig. 7.8) is of interest in relation to the formation of oxysulphide inclusions in steels.[29] Consider, for example, iron-rich alloys whose compositions (expressed in terms of oxygen–sulphur ratio) are higher than the line drawn from Fe through point *C* (Fig. 7.8*c*). In such alloys, e.g. composition *X*, the solidification process involves the liquid miscibility gap. For example, alloy *X* (say 0.1 wt% O, 0.2 wt% S) will commence solidification by the formation of δ-iron crystals. The liquid composition changes along the extension of the line joining Fe and *X*, i.e. it becomes enriched in oxygen and sulphur until its composition comes to lie on the monotectic curve at point *Y*. The monotectic reaction $L_1 \longrightarrow \text{Fe} + L_2$ commences as cooling proceeds (designated L_1 for liquid lying on the portion *AC* of the monotectic curve and L_2 for liquid lying on the *BC* portion of this curve). As the reaction proceeds the compositions of L_1 and L_2 move along the curves *AC* and *BC*, respectively. Under conditions of true equilibrium, L_1 is consumed; then L_2, on cooling, deposits iron while its composition changes on the extension of the line joining Fe and *X*. The composition of L_2 meets the eutectic valley $L \longrightarrow \text{Fe} + \text{FeO}$; this

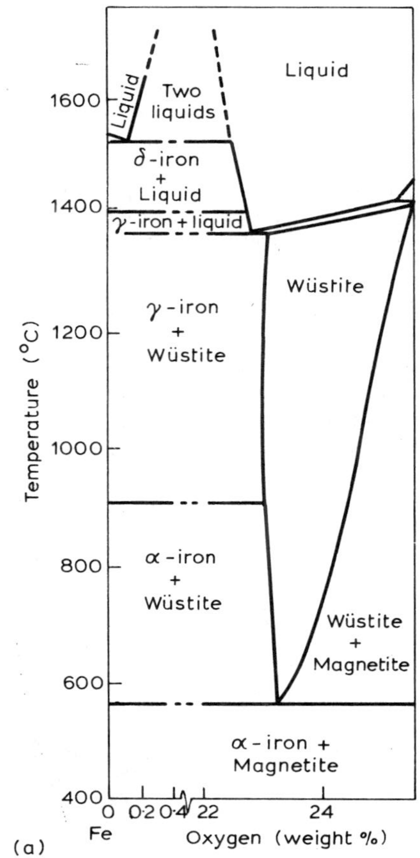

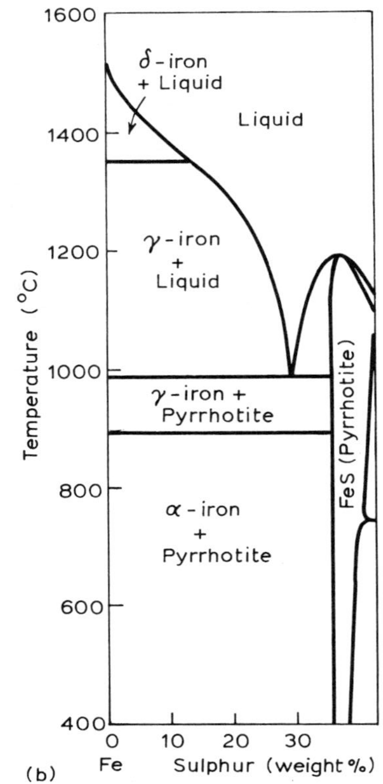

Fig. 7.8 (*a*) Part of the Fe–O binary system (*b*) Part of the Fe–S binary system

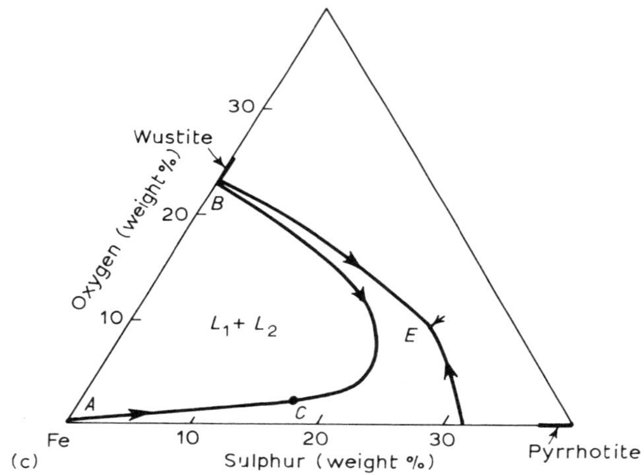

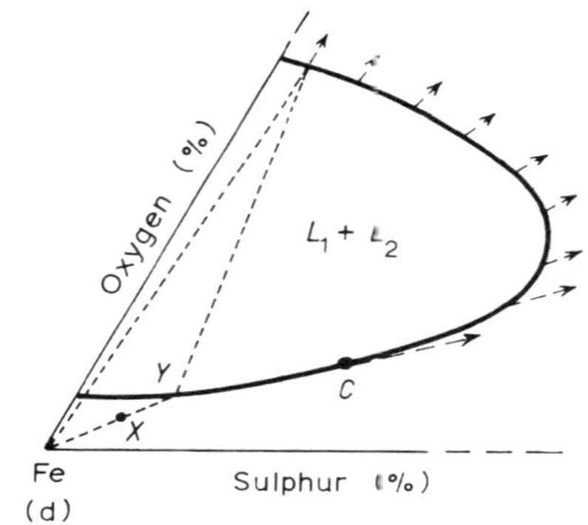

Fig. 7.8 (*cont.*)

(*c*) The liquidus projection for the iron-rich portion of the Fe–O–S system.[29] The binary Fe–O system contains a monotectic reaction: L_1 (~ 0·16 wt % O) ⇌ δ-Fe + L_2 (~ 22·6 wt % O) (1528°C) and this extends into the ternary system; the liquid miscibility gap is represented by the curve *ACB* where *C* is the 'critical point' occurring at a temperature minimum (~ 1340°C). Three binary eutectic reactions are shown: L_2 ⇌ γ-Fe + FeO (wustite) (1371°C); *L* ⇌ γ-Fe + FeS (pyrrhotite) (983°C) and *L* ⇌ FeO + FeS. The eutectic valleys meet at a ternary eutectic point *L* ⇌ α-Fe + FeO + FeS (920°C). The solubilities of oxygen and sulphur in solid iron are taken to be negligible in the diagram and the equilibria involving the various allotropes of iron are not shown

(*d*) Schematic view of the liquid miscibility gap (with enlargement of scale in the iron-rich corner) showing the solidification of alloy X under non-equilibrium conditions involving the formation of isolated droplets of L_2. The tie-triangle (- - - - - -) indicates the compositions of Fe, L_1 and L_2 at the beginning of the monotectic reaction. The dashed lines arrowed (— →) represent the changes of composition of individual liquid droplets as solidification proceeds by the separation of iron. The L_1 compositions corresponding to the various L_2 droplets lie along *YC* and are not shown

eutectic reaction proceeds until the liquid composition reaches the ternary eutectic point *E*, when the reaction $L \longrightarrow \text{Fe} + \text{FeO} + \text{FeS}$ completes the solidification.

In typical conditions of non-equilibrium solidification involving dendritic growth, the solidification process is modified significantly (Fig. 7.8*d*). When the formation of L_2 occurs, regions of this liquid become isolated by the growing dendrites, and no mass transport occurs between these regions. As solidification proceeds the composition of L_1 changes along curve *AC* while a series of droplets of L_2 form slightly richer in sulphur and of decreased oxygen/sulphur ratio, compared with the first formed droplets of L_2. In general, the compositions of L_1 and L_2 will move to point *C*, so that the compositions of the series of liquid droplets span the range up to *C*. With continued cooling each droplet of L_2 solidifies as an isolated region by further separation of iron. The composition of the liquid in each droplet changes on a path extending radially from the Fe corner of the system. These individual paths will intersect either the $L \longrightarrow \text{Fe} + \text{FeO}$ eutectic valley or the Fe + FeS eutectic valley (or for one particular composition will intersect the ternary eutectic point). Solidification of each droplet is completed by the Fe + FeO + FeS ternary eutectic reaction. Concerning liquid L_1, when its composition has reached point C, its solidification proceeds by the separation of iron, followed by the reactions $L \longrightarrow \text{Fe} + \text{FeS}$ and $L \longrightarrow \text{Fe} + \text{FeO} + \text{FeS}$. Thus, under these conditions of non-equilibrium solidification, the oxy-sulphide inclusions show a range of compositions and hence structural features which is in contrast to the uniform composition of inclusions that would form under equilibrium conditions.

Problems

1. By reference to schematic diagrams of tie-triangles, show how L_1 is consumed during the monotectic reaction in alloy *X*, assuming equilibrium conditions.

2. For alloys undergoing the monotectic reaction, assuming solidification conditions involving isolation of liquid regions, comment on how the overall alloy composition affects the amounts and the compositions of the inclusions.

Ag–Pb–Zn

The Ag–Pb–Zn system (Fig. 7.9) provides a basis for understanding the Parkes process and its various developments which use zinc additions to desilverize lead bullion. Industrial practice includes both batch processes

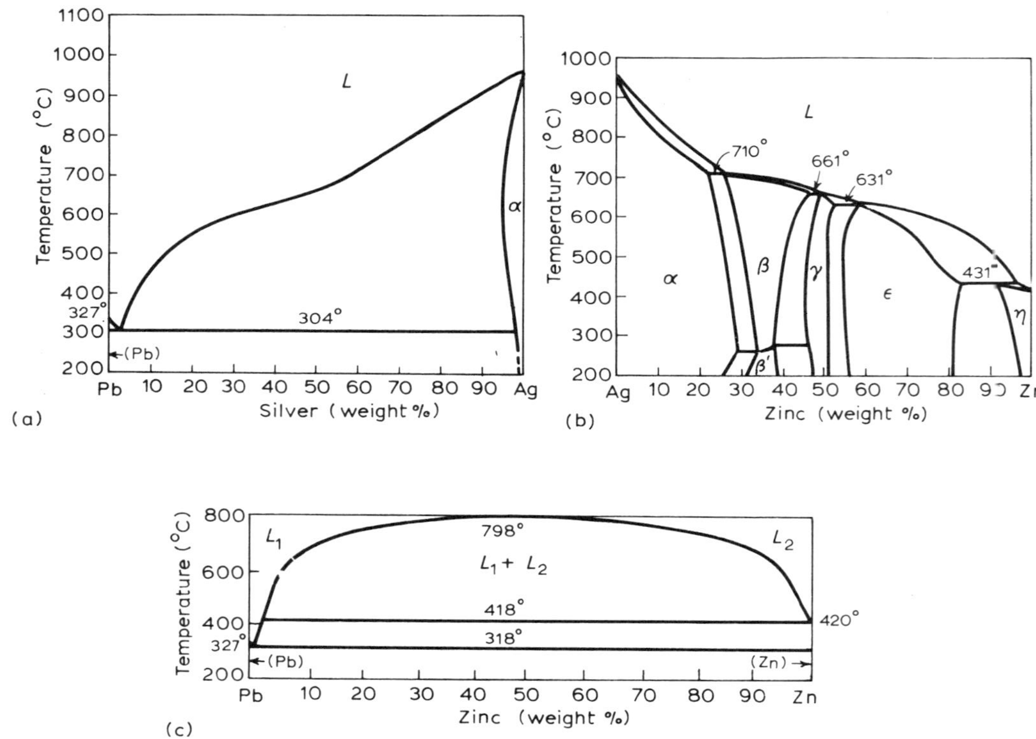

Fig. 7.9

(*a*) Pb–Ag binary system
(*b*) Ag–Zn binary system
(*c*) Pb–Zn binary system

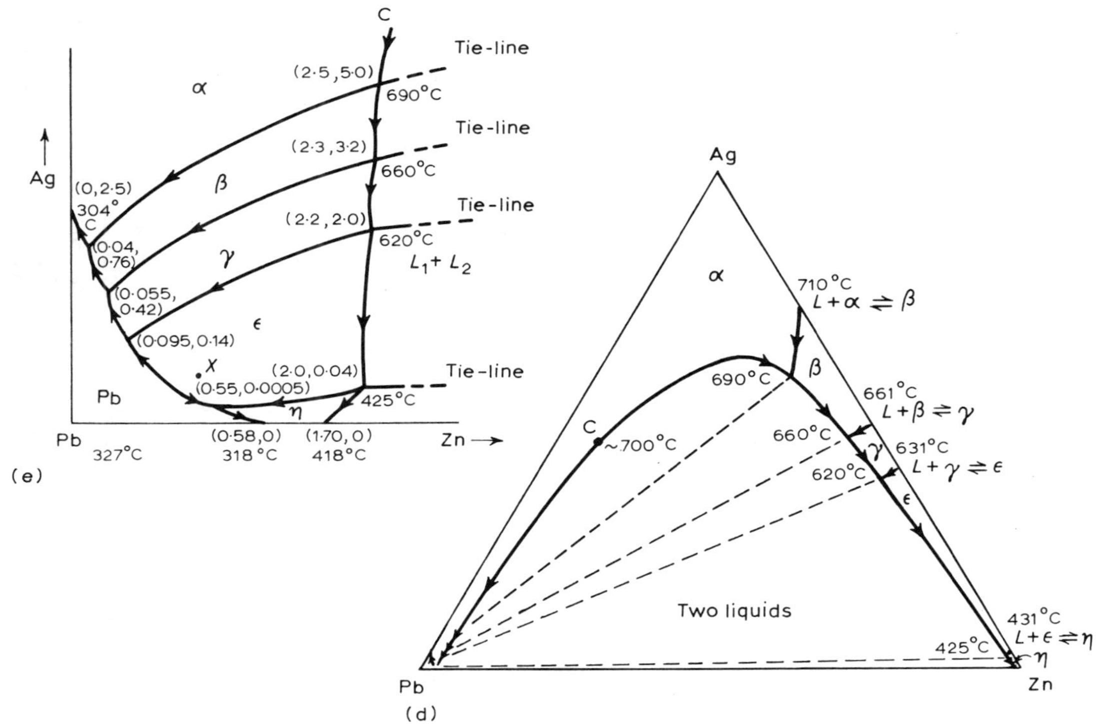

Fig. 7.9 (*cont.*)

(*d*) Liquidus projection of Ag–Pb–Zn system[30]

(*e*) The details (schematically) of the Pb-rich corner which are omitted in (*d*): compositions are in wt% and temperatures are from reference 30

(sometimes with several stages of zinc additions), and continuous processes.[30] The constituent binary systems are shown in Figs. 7.9*a–c*. The Ag–Pb system has a eutectic and there is very limited solid solubility. The Ag–Zn system contains primary solid solutions, α (based on Ag) and η (based on Zn) and three intermetallic compounds β, γ and ε which form peritectically. The Pb–Zn system contains a eutectic and there is very limited solid solubility; there is also a monotectic reaction. The liquid miscibility gap associated with the monotectic extends over a substantial part of the ternary system. The boundary of this gap (where the 'dome' intersects the liquidus surface) has a maximum temperature at ~700°C (point *C*) (Fig. 7.9*d*).

The four peritectic reaction curves extending from the Ag–Zn binary intersect the miscibility gap and give rise to invariant monotectic reactions: $L_1 + \alpha \leftrightharpoons L_2 + \beta$ (690°C); $L_1 + \beta \leftrightharpoons L_2 + \gamma$ (660°C); $L_1 + \gamma \leftrightharpoons L_2 + \varepsilon$ (620°C); $L_1 + \varepsilon \leftrightharpoons L_2 + \eta$ (425°C). Tie-lines join the compositions of the respective L_1 and L_2 points associated with these invariant reactions. The liquidus fields for primary separation of β, γ, ε and η are each divided into two regions by the liquid miscibility gap. These regions lie close to the Ag–Zn binary and Pb-rich corner, respectively.

In the Pb-rich corner of the system, Pb is involved in the following three-phase eutectic reactions (Fig. 7.9*e*): $L \leftrightharpoons \text{Pb} + \eta$; $L \leftrightharpoons \text{Pb} + \varepsilon$ (the reaction curve shows a temperature maximum); $L \leftrightharpoons \text{Pb} + \gamma$; $L \leftrightharpoons \text{Pb} + \beta$ $L \leftrightharpoons \text{Pb} + \alpha$. The Pb-rich corner also includes the three-phase reaction curves that emerge from the miscibility gap involving respectively: L, α, β; L, β, γ; L, γ, ε; L, ε, η. The curve for the composition of liquid in equilibrium with ε and η is peritectic along its whole length, i.e. $L + \varepsilon \leftrightharpoons \eta$; however, the other curves are eutectic over part of their length and then change to become peritectic in nature.[30] $L + \varepsilon \leftrightharpoons \gamma$; $L + \gamma \leftrightharpoons \beta$; $L + \beta \leftrightharpoons \alpha$. These transitions arise from changes in the liquid compositions relative to the solid phase compositions (see Fig. 3.20). The intersections of these curves with the eutectic curves involving Pb give rise to the following invariant reactions: $L + \varepsilon \leftrightharpoons \text{Pb} + \eta$; $L + \varepsilon \leftrightharpoons \text{Pb} + \gamma$; $L + \gamma \leftrightharpoons \text{Pb} + \beta$; $L + \beta \leftrightharpoons \text{Pb} + \text{Ag}$. (In the above summary of the Pb-rich corner, the subscript 2 has not been applied in designating the liquid phase.)

The desilverization process, which involves intermetallic compound separation from the liquid on cooling, can best be illustrated by reference to the enlarged view of the Pb-rich corner of the system shown in Fig. 7.9*e*. Zinc is added to molten lead (e.g. an addition of 2% Zn may be made to lead containing 0.2% Ag) and the melt is cooled to just above the melting point of lead. A solid 'crust' is formed, rich in zinc, which contains most of the silver (and gold), together with some lead. Liquid rich in lead precipitates the intermetallic Ag–Zn compound ε as cooling proceeds. The exact solidification sequence depends on the liquid composition, but to attain very low silver contents (e.g. <0.0005 %) the final stages of cooling must involve the separation of the zinc-rich solid solution, η. Thus, a sequence for alloy *X* in Fig. 7.9*e* would be:

$L \longrightarrow \varepsilon$
$L \longrightarrow \mathrm{Pb} + \varepsilon$
$L + \varepsilon \longrightarrow \mathrm{Pb} + \eta$
$L \longrightarrow \mathrm{Pb} + \eta.$

Whether or not two liquids form during a desilverization process depends on the amount of zinc added and the temperature. The Parkes process does not involve the formation of two liquids. However, the Williams continuous process uses a vessel with a temperature gradient from ~600–650°C at the top to near to the melting point of lead at the bottom. Lead, containing silver, is introduced at the top of the vessel, where a layer of molten zinc is maintained; about 10 % Zn can be dissolved by the lead. The precipitation of ε-AgZn occurs as the lead flows downwards through the vessel and is cooled, and desilverized lead is removed. The 'crust' is removed and is treated by distillation to recover the zinc, and by cupellation (selective oxidation) to remove the lead.

Problems

1. Draw, schematically, the part of the isothermal section of the Ag–Pb–Zn system at a temperature just above the melting point of lead (i.e. at ~330°C) showing the equilibria between lead-rich liquid and the Ag–Zn intermetallic compounds γ, ε and η.

2. What is the significance of the invariant reaction $L + \varepsilon \longrightarrow \mathrm{Pb} + \eta$ in relation to the degree of desilverization than can be achieved in practice?

Al_2O_3–MgO–SiO_2 (corundum–periclase–cristobalite/tridymite)

This ternary system (Fig. 7.10) is important in relation to various ceramics[31, 32] non-metallic inclusions in steels[33], slags[34] and also in petrology.[35, 36]

It contains binary and ternary compounds as follows:

$MgO{\cdot}Al_2O_3$ (spinel):	congruently melting	~2135°C
$3Al_2O_3{\cdot}2SiO_2$ (mullite):	congruently melting	~1850°C
$2MgO{\cdot}SiO_2$ (forsterite):	congruently melting	~1900°C
$MgO{\cdot}SiO_2$ (enstatite):	incongruently melting	1557°C

(There are three modifications of $MgO{\cdot}SiO_2$:

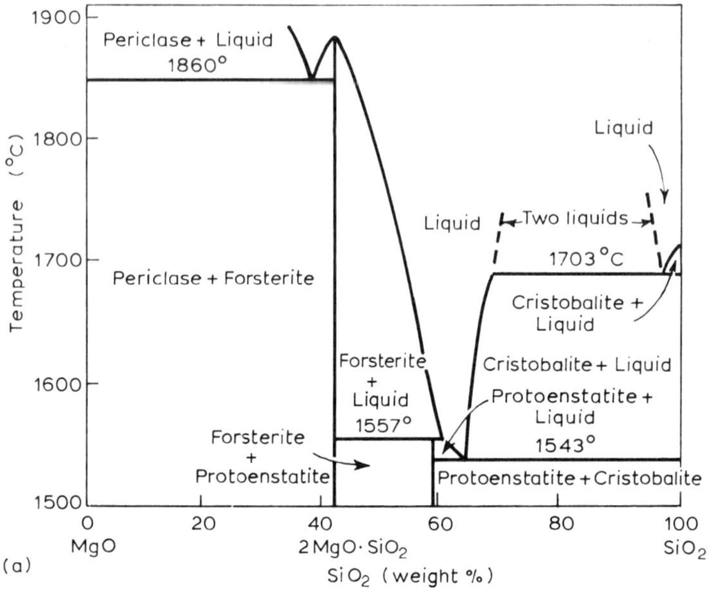

1900
1800
1700
1600
1500
Temperature (°C)
Periclase + Liquid
1860°
Liquid
Two liquids
Liquid
1703 °C
Periclase + Forsterite
Cristobalite + Liquid
Forsterite + Liquid
1557°
Cristobalite + Liquid
Protoenstatite + Liquid
1543°
Forsterite + Protoenstatite
Protoenstatite + Cristobalite
0
20
40
60
80
100
MgO
2 MgO · SiO2
SiO2
SiO2 (weight %)

(a)

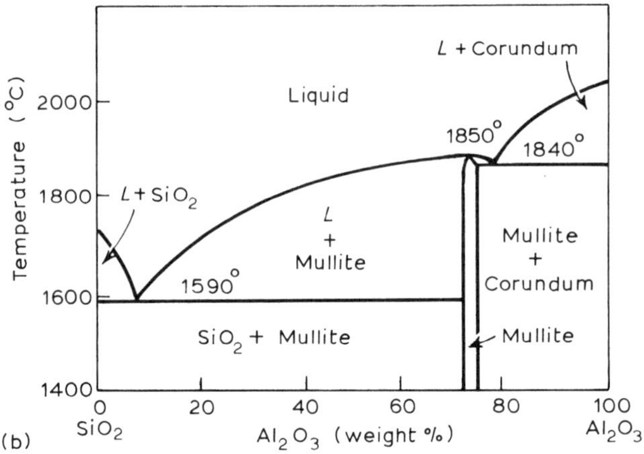

Temperature (°C)
2000
1800
1600
1400
L + Corundum
Liquid
1850°
1840°
L + SiO2
L + Mullite
Mullite + Corundum
1590°
SiO2 + Mullite
Mullite
0
20
40
60
80
100
SiO2
Al2O3 (weight %)
Al2O3

(b)

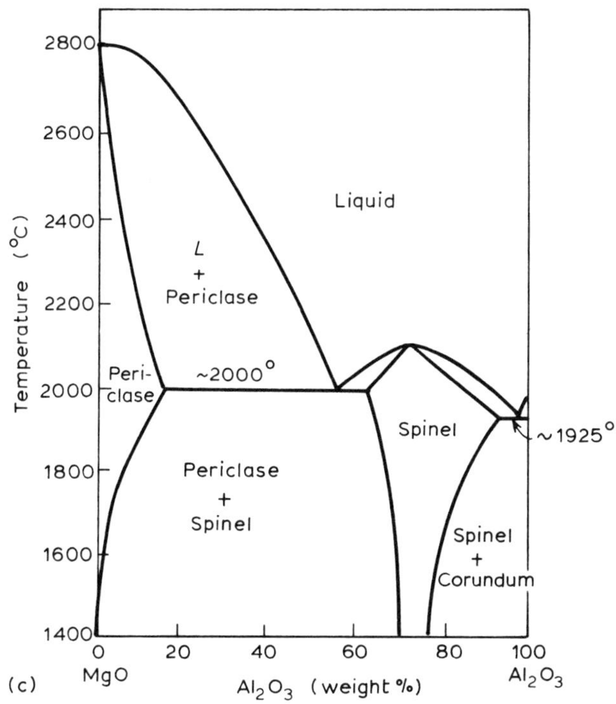

Fig. 7.10 (*a*) MgO–SiO_2 binary system

(*b*) SiO_2–Al_2O_3 binary system

(*c*) MgO–Al_2O_3 binary system. (Note: the value of ~ 2000°C shown for the temperature of the $L \rightleftharpoons$ periclase + spinel reaction is taken from a more recent source[32] than that used for the value of ~ 1850°C shown in (*d*)[37])

(*d*) Composite representation of Al_2O_3–MgO–SiO_2 system, showing liquidus projection and boundaries of phase regions in the solid state. The latter include information (hatched lines) on ranges of solubility for spinel ($MgO \cdot Al_2O_3$), mullite ($3Al_2O_3 \cdot 2SiO_2$) and cordierite ($2MgO \cdot 2Al_2O_3 \cdot 5SiO_2$); however solid solution regions are not shown for MgO or for the MgO-rich range of spinel. The limits of the solid solubility ranges are markedly temperature dependent.[37] Note: A three-dimensional representation (anaglyph) of the Al_2O_3–MgO–SiO_2 system is given in Reference 38

(*e*) Isothermal section of Al_2O_3–MgO–SiO_2 system at 1460°C.[31, 36] The range of solid solutions for cordierite is shown as a heavy line but solid solution ranges for other compounds are not represented. The invariant reaction (at point *N*) liquid (b) + mullite $\rightleftharpoons$ sapphirine + cordierite occurs at 1460°C and is represented by the four-phase region liquid (b) + mullite + sapphirine + cordierite; the dashed lines within this region connect the pairs of phases involved. Abbreviations used: cordierite (Co), forsterite (Fo), mullite (Mo), protoenstatite (PE), sapphirine (Sa), spinel (Sp), liquid (Liq)

(*d* and *e* reproduced by permission of the American Ceramic Society Inc. from references 31 and 37) (*d*) and (*e*) appear on following pages

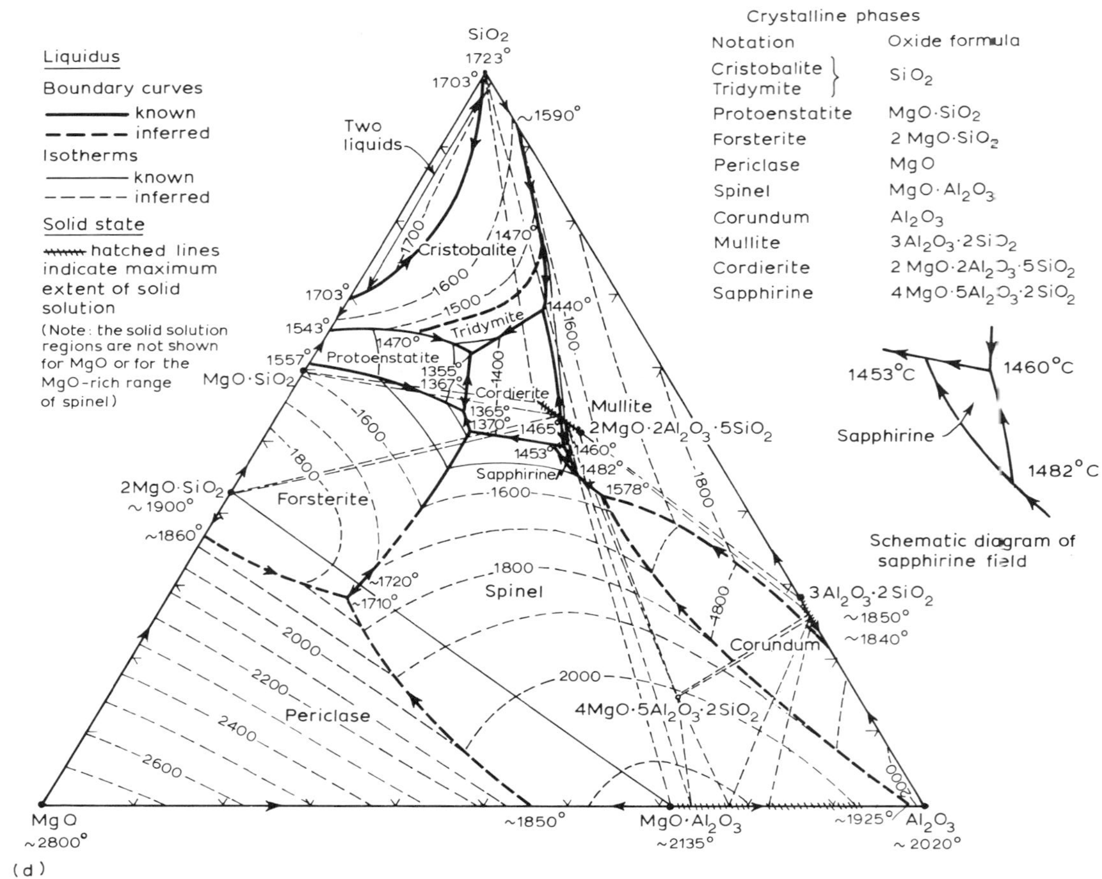

Liquidus
Boundary curves
known
inferred
Isotherms
known
inferred
Solid state
hatched lines indicate maximum extent of solid solution
(Note: the solid solution regions are not shown for MgO or for the MgO-rich range of spinel)
SiO2
1723°
1703°
~1590°
Two liquids
Cristobalite
Tridymite
Protoenstatite
Cordierite
Mullite
Sapphirine
Forsterite
Spinel
Periclase
Corundum
MgO·SiO2
2MgO·SiO2
~1900°
~1860°
1703°
1543°
1557°
1470°
1440°
1355°
1367°
1365°
1370°
1465°
1460°
1453°
1482°
1578°
~1720°
~1710°
2MgO·2Al2O3·5SiO2
3Al2O3·2SiO2
~1850°
~1840°
4MgO·5Al2O3·2SiO2
MgO
~2800°
~1850°
MgO·Al2O3
~2135°
~1925°
Al2O3
~2020°
1400
1500
1600
1700
1800
2000
2200
2400
2600
Crystalline phases
Notation
Oxide formula
Cristobalite
Tridymite
SiO2
Protoenstatite
MgO·SiO2
Forsterite
2MgO·SiO2
Periclase
MgO
Spinel
MgO·Al2O3
Corundum
Al2O3
Mullite
3Al2O3·2SiO2
Cordierite
2MgO·2Al2O3·5SiO2
Sapphirine
4MgO·5Al2O3·2SiO2
1453°C
1460°C
1482°C
Sapphirine
Schematic diagram of sapphirine field
(d)

Fig. 7.10 (*cont.*)

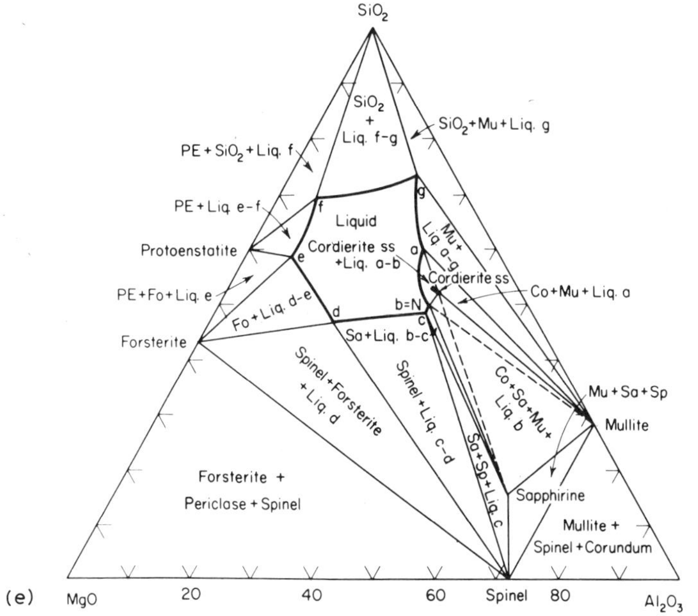

Fig. 7.10 (*cont.*)

protoenstatite (the stable high temperature form >1040°C);
enstatite (the stable low temperature form);
clinoenstatite (can exist as a metastable form at room temperature))

$4MgO{\cdot}5Al_2O_3{\cdot}2SiO_2$ (sapphirine): incongruently melting 1482°C

$2MgO{\cdot}2Al_2O_3{\cdot}5SiO_2$ (cordierite): incongruently melting 1465°C

Of these compounds spinel, mullite and cordierite show ranges of solid solution at certain temperatures, that of spinel being particularly extensive, (from ~60–80% Al_2O_3). These maximum ranges are shown by the hatched lines in Fig. 7.3*a* and there are corresponding extensive two-phase regions in the solid state. Other features to note in the system include the liquid miscibility gap based on the MgO–SiO_2 system and the existence of cristobalite and tridymite as allotropic forms of SiO_2.

The Al_2O_3–MgO–SiO_2 system is an example of relatively stable oxides containing cations which appear in only one predominant state of oxidation. In this system the gas phase can be ignored and the total content of condensed phases remains effectively constant. For an example of a ternary system where the gas phase is important reference may be made to the Fe–O–Si system[34]; solidification can be considered either under conditions of constant total composition of condensed phases or at constant oxygen pressure.

Invariant reactions are as follows:

Binary systems (Figs 7.10*a–c*)

$MgO–Al_2O_3$	~1850°C	$L \leftrightharpoons$ MgO (periclase) + spinel
	~1925°C	$L \leftrightharpoons Al_2O_3$ (corundum) + spinel
$Al_2O_3–SiO_2$	~1840°C	$L \leftrightharpoons Al_2O_3$ + mullite
	~1590°C	$L \leftrightharpoons SiO_2$ (cristobalite) + mullite
$MgO–SiO_2$	~1860°C	$L \leftrightharpoons$ MgO (periclase) + forsterite
	1557°C	L + forsterite $\leftrightharpoons$ protoenstatite
	1543°C	$L \leftrightharpoons$ protoenstatite + SiO_2 (cristobalite)
	1703°C	$L_1 \leftrightharpoons L_2 + SiO_2$ (cristobalite)

Ternary compositions

1578°C	$L + Al_2O_3$ (corundum) $\leftrightharpoons$ spinel + mullite
1482°C	L + spinel + mullite $\leftrightharpoons$ sapphirine
1460°C	L + mullite $\leftrightharpoons$ sapphirine + cordierite
1453°C	L + sapphirine $\leftrightharpoons$ spinel + cordierite
1440°C	L + mullite $\leftrightharpoons$ cordierite + SiO_2 (tridymite)
1355°C	$L \leftrightharpoons SiO_2$ (tridymite) + cordierite + protoenstatite
1370°C	L + spinel $\leftrightharpoons$ forsterite + cordierite
1365°C	$L \leftrightharpoons$ forsterite + protoenstatite + cordierite
1710°C	$L \leftrightharpoons$ MgO (periclase) + forsterite + spinel

Note also the 'saddle' points

1720°C	$L \leftrightharpoons$ forsterite + spinel
1367°C	$L \leftrightharpoons$ protoenstatite + cordierite
1465°C	L + mullite $\leftrightharpoons$ cordierite

Note on Terminology

Having considered here a detailed example of a ceramic system, it is appropriate to mention certain features of the terminology which are sometimes applied to phase diagrams in ceramics and in geology[31, 35] and which differ from the usual metallurgical terminology.

Curves representing the compositions of liquid in equilibrium with two other phases and which subdivide the liquidus into primary fields may be called *boundary curves.* With reference to solidification they may be described as either *subtraction curves*, along which crystalline phases separate from the liquid (corresponding to eutectic solidification, e.g. $L \longrightarrow \alpha + \beta$) or *reaction curves*, along which crystalline phases both precipitate and dissolve (peritectic solidification, e.g. $L + \alpha \longrightarrow \beta$).[35]

The term *Alkemade line* (or compatibility line) refers to a straight line joining the composition points of two co-existing phases whose primary phase fields share a common boundary curve. For example, in Fig. 5.2

the primary fields of *A* and *C* share the boundary curve *ji*, so that *AC*, the side of the composition triangle is an Alkemade line; the primary fields of *X* and *C* share the boundary curve *ihg* so that *XC* (the quasi-binary section) is an Alkemade line. This system (Fig. 5.2) contains five boundary curves and there are correspondingly five Alkemade lines *AX*, *XB*, *BC*, *AC*, *XC*. A liquid whose original composition lies on an Alkemade line will solidify to a mixture of the two solid phases whose compositions lie at the ends of the line.

Alkemade's theorem states that the intersection of an Alkemade line with its pertinent boundary curve represents a temperature maximum for the boundary curve, e.g. in Fig. 5.2. points *d*, *e*, *f*, *j*, *h*. If the boundary curve (or any tangent drawn from it) intersects an extension of the pertinent Alkemade line, but not the actual line then that part of the boundary curve is of the reaction type. The point of intersection again represents a temperature maximum for the boundary curve, e.g. in Fig. 5.11 a tangent drawn to reaction curve *dh* intersects the extension of line *AX*.

Concerning ternary invariant reactions, the types designated peritectic (ii) and (iii) on p.86 can be described in terms of the points where the reaction curves intersect as *tributary reaction points* and *distributary reaction points* respectively.[35] For each invariant reaction there is an *Alkemade triangle* (compatibility triangle), formed from three Alkemade lines.

The following procedure may be followed in approaching the interpretetion of a complex ceramic system. Identify the Alkemade lines (one for each boundary curve). Find the nature of the boundary curves by determining whether the tangents intersect the pertinent Alkemade line, or its extension (cf. Fig. 3.17). Taking into account the slopes of the boundary curves identify the nature of the reaction points (p.86). The application of this approach to complex systems such as Al_2O_3–MgO–SiO_2 and Al_2O_3–CaO–SiO_2 is described by Ehlers.[35]

Problems

1. Draw an isothermal section representing the solid state constitution for the Al_2O_3–MgO–SiO_2 system.

2. By reference to Fig. 7.10*d*, describe the solidification sequences of the following ceramic compositions (wt %), assuming equilibrium conditions.

(a) 15MgO–5SiO_2 consisting of spinel + mullite when solidification is completed.

(b) 22MgO–7SiO_2 consisting of spinel + sapphirine when solidification is completed.

(c) Sapphirine composition consisting of sapphirine only when solidification is completed.

(d) 8MgO–17Al_2O_3 consisting of SiO_2 + cordierite when solidification is completed.

3. Compare the melting characteristics of the following ceramic compositions (i.e. the variation of liquid content with temperature within the melting range) and comment on the relevance to the firing behaviour.

(a) 45% SiO_2–4% Al_2O_3 (forsterite type)
(b) 65% SiO_2–5% Al_2O_3 (steatite type)
(c) 58% SiO_2–24% Al_2O_3 (cordierite type)

4. In the context of the fluxing effect of MgO on Al_2O_3–SiO_2 fire-clay bricks, describe the result of adding 1% MgO to a material of composition 42% Al_2O_3, 58% SiO_2, heated to equilibrium at 1460°C (refer to Fig. 7.10*e*).

Answers to Problems

The use of triangular coordinate graph paper (e.g. Chartwell 5590) is recommended for questions involving the plotting of composition data. Alternatively, page 149 can be photocopied to provide graph paper.

Chapter 1

(a) *Procedure:* Plot the compositions of the three phases to form a tie-triangle and also plot the alloy composition. Measure the appropriate lengths from this diagram (e.g. cf. tx_1 and xx_1 in Fig. 1.6) for the application of the 'centre of gravity' principle.

Answer: ~55% Liquid, 10%α, 35%γ.

(b) *Procedure:* From the lever rule the mid-point of the α–γ side of the tie-triangle is the composition required.

Answer: ~48%A, 15%B, 37%C.

Chapter 2

Procedure: Draw possible forms of the binary diagrams assuming the liquidus and solidus curves to be smooth. Then, for diagrams (a) and (b) locate the binary alloy compositions corresponding to the quoted temperatures, e.g. for a liquidus of 950°C in binary system AB the alloy composition is ~70% B, 30% A, while the same liquidus in AC corresponds to an alloy composition of ~35% C, 65% A. On the ternary composition plot, draw a smooth curve, joining these two binary compositions, to represent the possible form of the liquidus isotherm (cf. Fig. 2.2). Draw a similar construction joining the solidus compositions for 950°C. The isothermal and vertical sections can then be constructed from diagrams (a) and (b) (see Figs. 2.3 and 2.4).

Answer: See Figs. A1*a* and *b*.

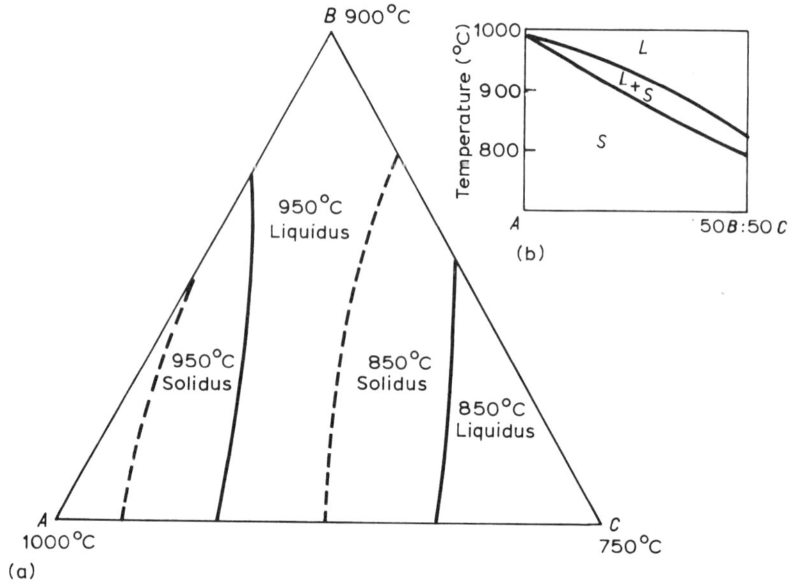

Fig. A1. (*a*) Liquidus and solidus projections showing isotherms for 950°C and 850°C (*b*) Vertical section between *A* and the midpoint of *BC*

Chapter 3

1. (a) (i) $\alpha/\alpha + L$; (ii) $L/L + \beta$; (iii) $\alpha + L/\alpha + \beta + L$; (iv) $\alpha + \beta + L/\alpha + \beta$. (b) (i) *BEF*; (ii) *MEFN*.

2. *Procedure*: Use the given data to plot tie-triangles for 540°C and 510°C, respectively, and calculate the proportions of the phases using the 'centre of gravity' principle.

Answer: (a) ~36%α, 27%β. (b) Liquid at 540°C/liquid at 510°C ~2.9.

3. (a) $L \longrightarrow$ primary β, commencing at temperature t_l; $L \longrightarrow \alpha + \beta$ eutectic reaction, commencing at temperature t_e; see Figs. 3.1 and 3.3 for the compositional changes involving $L + \alpha + \beta$ tie triangles: the eutectic reaction is completed at temperature t_s to give an $\alpha + \beta$ structure. (b) *MEFN*; *DEFG*.

4. *Procedure*: (a) From the data given (temperatures of the invariant reactions and the compositions of the phases involved) it follows that the binary invariant reactions are of the eutectic type $L \longrightarrow \alpha + \beta$. Draw a smooth curve joining the compositions of the eutectic liquid in systems

AB and AC, to represent the liquidus projection. The solidus projection is similarly obtained by joining respectively the α compositions in AB and AC and the β compositions (cf. Fig. 3.4).

(b) The alloy containing 30% A, 55% B and 15% C, commences its solidification by the formation of the primary β phase. Solidification is completed by the eutectic reaction $L \longrightarrow \alpha + \beta$ at a temperature when the $\alpha + \beta$ tie-line passes through the alloy composition point (see Fig. 3.3). The position of this tie-line cannot be deduced accurately from the given data, but assuming a possible position for this line the lever-rule can be used to calculate the proportions of α and β.

Answer: (b) ~23% α, 77% β.

5. (a) (i) $L + \alpha$; (ii) $L + \beta$; (iii) $L + \alpha + \beta$. (b) (i) *EDGF*; (ii) *AMN*.

6. *Procedure*: (a) Plot the data in the form of three tie-triangles, representing the equilibria between liquid, α and β at 550, 520 and 500°C respectively.

Apply the 'centre-of gravity' principle to the particular alloy quoted in relation to the 520°C tie-triangle.

Answer: (a) ~18% *L*, 45% α, 37% β.

(b) The data for the liquid, α and β corners of the tie-triangles enable parts of the liquidus and solidus projections to be represented, e.g. by drawing a curve through the liquid compositions for the equilibria at 550°C, 520°C and 500°C respectively.

Draw a curve through the corners of the tie-triangles representing the liquid compositions and thus obtain a part of the liquidus projection. Proceed similarly for the α and β corners of the tie-triangles to obtain parts of the solidus projection. Then applying the 'tangent' procedure illustrated in Fig. 3.17, it can be deduced that the three-phase reaction is a peritectic $L + \beta \longrightarrow \alpha$.

(c) From the fact that only three phases are present in the system, namely liquid, α and β, respectively, it follows that one of the binary systems shows complete solid solubility. Phase Rule considerations indicate that no invariant reactions occur within the ternary system, and if it is assumed that both systems AB and BC contain a peritectic reaction and system AC shows complete solid solubility then the liquidus projection will consist of a curve joining the peritectic liquid composition values in systems AB and BC (assume reasonable values for these compositions). The solidus projection will consist of two curves joining the α and β compositions involved in the binary peritectic reactions (assume reasonable values for these). The general form of the liquidus and solidus projections will be as in Fig. 3.14.

To show an isothermal section in the solid state neglect the change of solid solubility with temperature, and the solidus projection may then be taken to correspond to the required isothermal section. Alternatively, assume that some decrease in the solubility of A and C in β, and of B in α occurs, and draw a possible isothermal section with boundaries displaced from the solidus projection. The isothermal section for 550°C will incorporate the 550°C tie-triangle composition data; Fig. 3.15 provides a guide to the arrangement of the other regions in this section.

(d) Consideration of the alloy composition in relation to the liquidus projection and the solid state isothermal section shows that solidification commences with the separation of primary β, whereas the equilibrium structure of the solid is single phase α. Thus, the solidification sequence involves the peritectic reaction and the consumption of all the β phase during this reaction (see Fig. 3.13 and page 27). During the separation of primary β the liquid composition moves in a curved path on the liquidus surface, while the composition changes over the solidus surface; the precise paths cannot be deduced from the data given. At the end of the primary solidification the liquid + β tie-line forms one of the sides of a liquid + α + β tie-triangle. Consideration of the tie-triangle data shows that this stage is reached at a temperature close to 550°C. With falling temperature the $L + \alpha \longrightarrow \beta$ reaction proceeds and the tie-triangles indicate its progress (cf. Figs. 3.12 and 3.13). At the end of the reaction the alloy composition point lies on the $L + \alpha$ side of a tie-triangle and from the tie triangle data it can be seen that this stage is reached at ~520°C. The liquid composition now moves away, with falling temperature, from the peritectic liquid curve. As the composition moves over the α liquidus surface, α phase separates until the solidus is reached.

7. *Procedure*: Draw a line from the binary alloy composition 55Pb–45Zn parallel to the Zn–Sn side of the system. Determine the Sn contents corresponding to the intersections of this line with each of the tie-lines in the miscibility gap. Apply the lever-rule to each of the tie-lines to calculate the proportion of Pb-rich liquid.

Answer: The proportion of Pb-rich liquid varies from ~60% in the binary Pb–Zn alloy to close to 100% in the alloy containing 55Pb 20Zn 25Sn.

Chapter 4

1. (a) (i) $L + \beta$; (ii) $L + \alpha + \gamma$. (b) (i) M_1DGM; (ii) CO_1OO_2.

2. (a) (i) M_1N_1NM, N_1DGN, M_1DGM, MNG; (ii) ADM_1, $ADGE$, AM_1MM_2, M_1MGD, M_2MGE, AEM_2. (b) (i) $L + \gamma / L + \beta + \gamma$; (ii) $L + \alpha + \beta / L + \alpha + \gamma / \alpha + \beta + \gamma$; (iii) $L + \alpha + \beta / L + \alpha + \gamma / L + \beta + \gamma$.

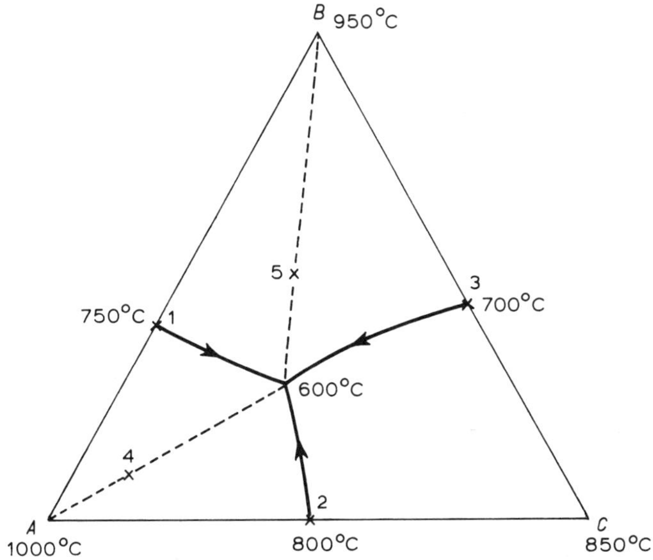

Fig. A2. Liquidus projection. Alloy compositions are marked x and numbered

3. *Procedure*: (See Fig. A2) (a) From the data, including negligible solid solubility and the absence of compounds, it can be deduced that the binary systems are of the eutectic type and that a ternary eutectic reaction $L \longrightarrow A + B + C$ occurs at 600°C. The data for alloys 1, 2 and 3 show the compositions and temperatures of the binary eutectic reactions: $L \longrightarrow A + B$; $L \longrightarrow A + C$ and $L \longrightarrow B + C$ respectively. To determine the ternary eutectic composition draw a line joining A to the composition of alloy 4; also draw a line joining B and the alloy 5 composition. Extend these lines to intersect one another; their point of intersection represents the ternary eutectic composition.

(b) The alloy composition lies on the line joining A to the ternary eutectic composition; apply the lever-rule to this line to calculate the proportion of primary A.

Answer: (a) See Fig. A2. (b) ~65% of primary A.

4. (a) (i) A liquidus temperature of ~770°C may be deduced by 'linear' interpolation between the 750°C and 800°C isothermal curves. A better procedure is to extrapolate using a construction of part of a vertical section passing through Cu and the alloy composition point and extending

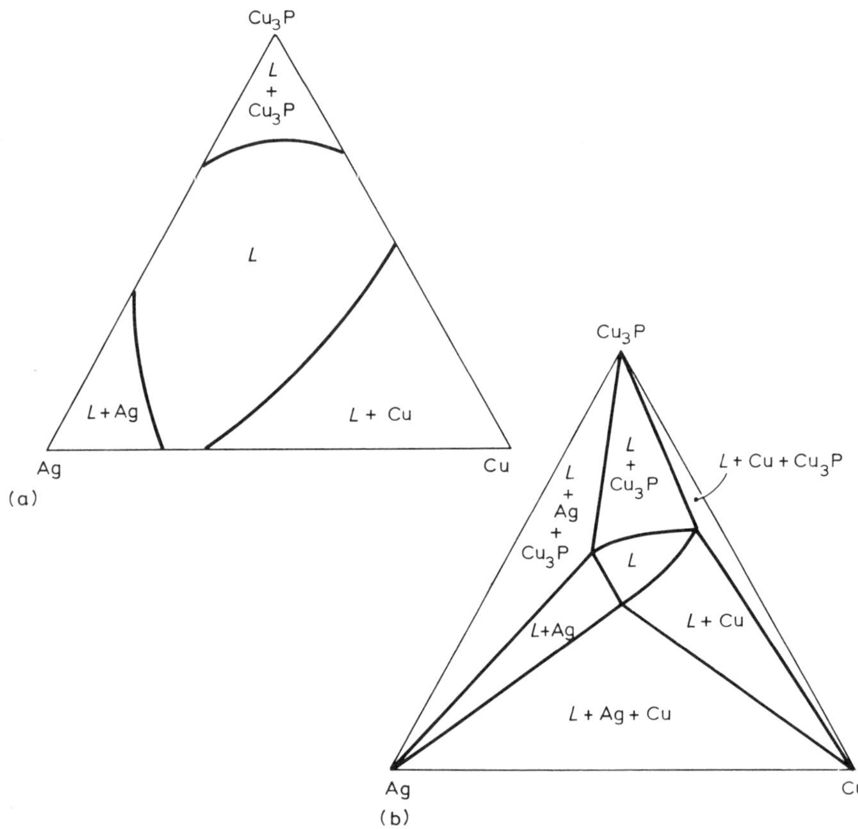

Fig. A3. Ag–Cu–Cu_3P system. Isothermal sections at (*a*) 800°C and (*b*) a temperature above the ternary eutectic but below the binary eutectics

as far as the intersection with the $L \longrightarrow Cu + Cu_3P$ curve. Values of liquidus temperatures can be plotted corresponding to the intersections of this section with the isotherms; distances to represent the composition scale are measured from Fig. 4.17 along the section from the Cu corner.

(ii)–(iv) Using the procedure described on pages 47 and 48:
(ii) ~87% liquid; (iii) ~40% liquid, ~30% Cu, ~30% Cu_3P; (iv) ~13% primary Cu, ~65% binary eutectic (Cu + Cu_3P) ~22% ternary eutectic (Ag + Cu + Cu_3P). Note that the proportion of binary eutectic can be obtained by difference, after the primary Cu has been calculated from the appropriate tie-line and the ternary eutectic from the appropriate tie-triangle.

(b) See Fig. A3.

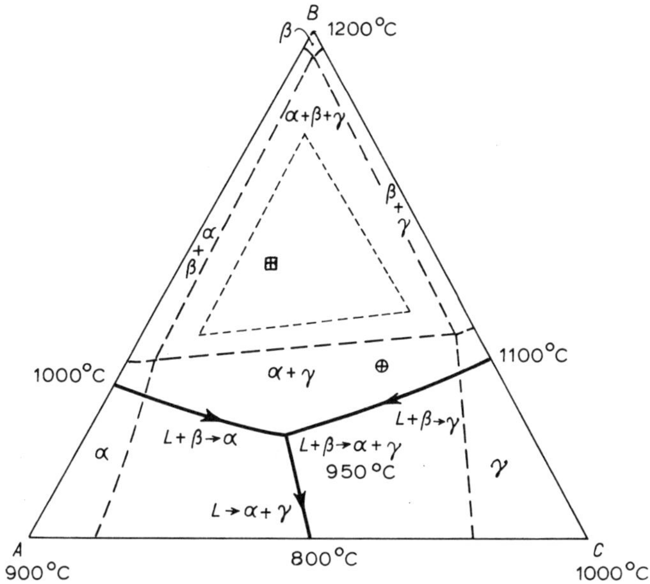

Fig. A4. ——— Liquidus projection
— — Isothermal section at room temperature
- - - - - Boundaries of $\alpha + \beta + \gamma$ region at 945°C

5. (a) See Fig. A4. (b) System AB, $L + \beta \longrightarrow \alpha$; BC, $L + \beta \longrightarrow \gamma$; AC, $L \longrightarrow \alpha + \gamma$; ABC, $L + \beta \longrightarrow \alpha + \gamma$. (c) Alloy ⊞ $L \longrightarrow \beta$; $L + \beta \longrightarrow \alpha$; $L + \beta \longrightarrow \alpha + \gamma$; $\alpha + \beta + \gamma$. By application of the centre of gravity rule to the $\alpha + \beta + \gamma$ tie-triangle at 945°C: ~48%α ~36%β ~16%γ. (d) Alloy ⊕ $L \longrightarrow \beta$; $L + \beta \longrightarrow \gamma$; β consumed; $L \longrightarrow \gamma$; $L \longrightarrow \alpha + \gamma$; $\alpha + \gamma$.

Chapter 5

1. (a) (i) and (ii) See Fig. A5. Invariant reactions. System AB, $L \longrightarrow \alpha + \beta$; BC, $L + \beta \longrightarrow \delta$; $L \longrightarrow \delta + \gamma$; AC, $L \longrightarrow \alpha + \gamma$; ABC, $L + \beta \longrightarrow \alpha + \delta$; $L \longrightarrow \alpha + \delta + \gamma$.

(b) ⊕ $L \longrightarrow \beta$; $L + \beta \longrightarrow \delta$; $L + \beta \longrightarrow \alpha + \gamma$; liquid consumed; final structure $\alpha + \beta + \delta$. For a temperature of 651°C (i.e. just above the invariant reaction plane $L + \beta \longrightarrow \alpha + \gamma$) apply the centre of gravity principle to the tie-triangle formed by joining the compositions of the peritectic liquid and the β and δ compositions. (Answer ~24% β). For 649°C use the α, β, δ tie-triangle. (Answer ~15% β).

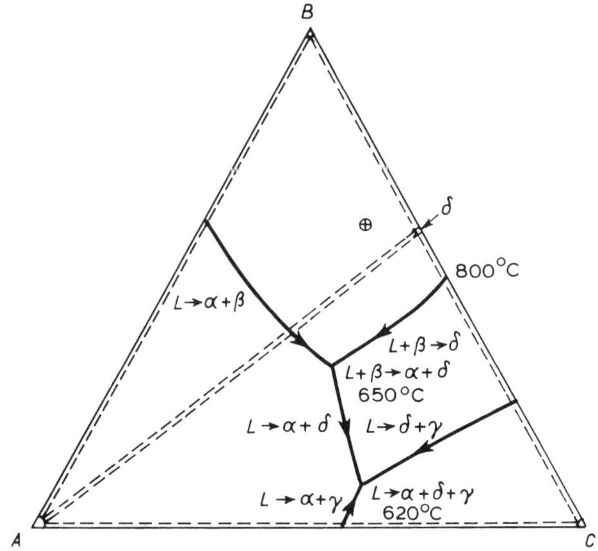

Fig. A5. ——— Liquidus projection
– – – – Isothermal section at 600°C containing three-phase regions $\alpha + \beta + \delta$ and $\alpha + \delta + \gamma$ with the associated single-phase and two-phase regions

2. (a) $L \longrightarrow \alpha$; $L + \alpha \longrightarrow \beta$ (along PX); $L \longrightarrow \alpha + \beta + \delta$; final structure $\alpha + \beta + \delta$. (Note: mention is made on pages 50, 74, 75 of the transition of a three-phase reaction from peritectic to eutectic to give rise to a ternary invariant eutectic reaction by intersection with two other eutectic curves. It is not necessary for the peritectic to show this transition in order to produce a ternary invariant eutectic; the reaction can remain peritectic in nature along the whole length of the liquid composition curve and this is the case for curve PX, representing $L + \alpha \longrightarrow \beta$ in Fig. 5.15).

(b) $L \longrightarrow \beta$; $L \longrightarrow \beta + \delta$ (along MX); final structure $\beta + \delta$.

3. (a) (i) Liquidus ~570°C; $L \longrightarrow$ Al; $L \longrightarrow$ Al + T; $L \longrightarrow$ Al + β + T at 450°C; final structure Al + β + T; (ii) Liquidus ~580°C; $L \longrightarrow$ Al; $L \longrightarrow$ Al + S; L + Al $\longrightarrow$ S + T at 467°C; liquid consumed; final structure Al + S + T. (iii) Liquidus ~520°C; $L \longrightarrow$ Al; $L \longrightarrow$ Al + S; final structure Al + S. (iv) Liquidus ~635° C; $L \longrightarrow$ Al; $L \longrightarrow$ Al + S; solidus ~515°C; final structure Al + S. Note: in each of the above cases the primary phase (designated Al) is a solid solution of Cu and Mg in Al; the exact compositions of the solid solutions that form cannot be deduced from the data provided and hence the exact paths of the liquid composition changes during primary solidification are not known.

(b) For the binary Al–4·5% Cu alloy, the structure at 500°C is single phase Al–solid solution, while for the ternary alloy it is Al solid solution + S phase. The solid solubility limit at 500°C is not given, but from the solidus projection it is seen that the Cu content of the solid solution in the ternary alloy is significantly lower than that in the binary alloy. On reheating the quenched alloys at 200°C, the structure of the binary becomes Al solid solution + θ, while in the ternary alloy additional S phase will precipitate.

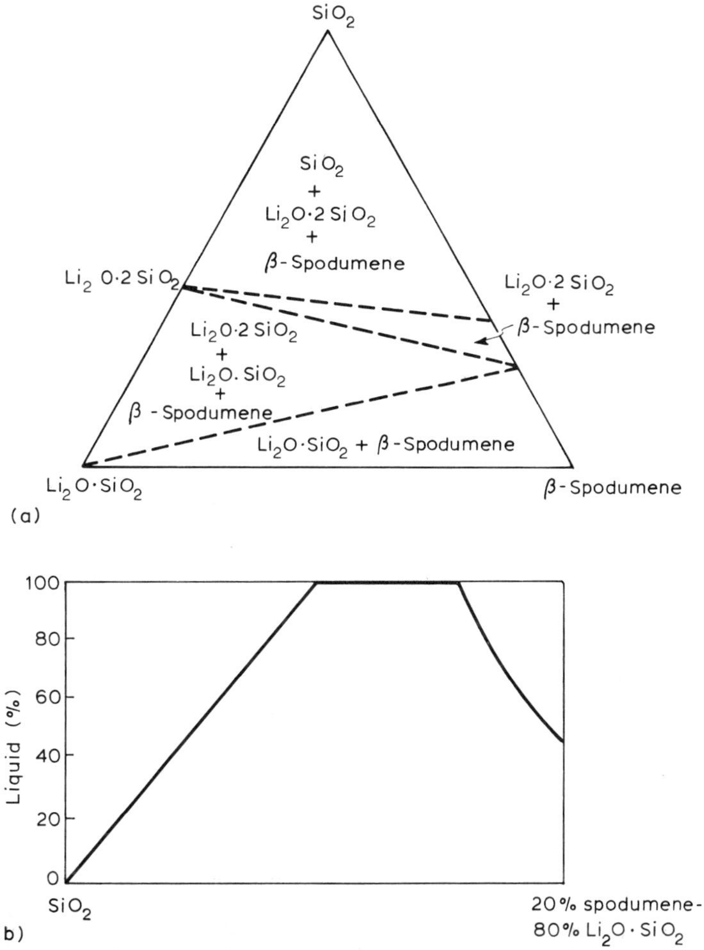

Fig. A6. (*a*) Isothermal section of the SiO_2–$Li_2O{\cdot}SiO_2$–$Li_2O{\cdot}Al_2O_3{\cdot}4SiO_2$ system at 970°C

(*b*) Variation of percentage liquid with composition at 1100°C along a line joining SiO_2 to 20% spodumene–80% $Li_2O{\cdot}SiO_2$

4. (a) See Fig. A6*a*.

(b) Consider the intersections of the composition line with the isotherms on the liquidus surface (Fig. 5.19*b*). From the SiO_2 end of the line the proportion of liquid increases linearly along the line up to 100% liquid, where the line intersects the 1100°C isotherm. Between this point and the intersection with the 1100°C isotherm in the primary $Li_2O{\cdot}SiO_2$ region, there is 100% liquid at 1100°C (Fig. A6*b*). For the remaining part of the composition line select various compositions; then obtain the proportions of liquid by applying the lever rule to the tie-lines (for each point selected) drawn from $Li_2O{\cdot}SiO_2$ to the corresponding point on the 1100°C isotherm.

Chapter 7

C–Nb–Ta

See for guidance Figs. 3.8, 3.16 and 5.12.

Answer: Fig. A7 (schematic). The alloy lies in the $M_2C + MC$ region; incongruent melting of M_2C occurs by the reaction $M_2C \longrightarrow L + MC$; melting continues with increase in temperature through the $L + MC$ region.

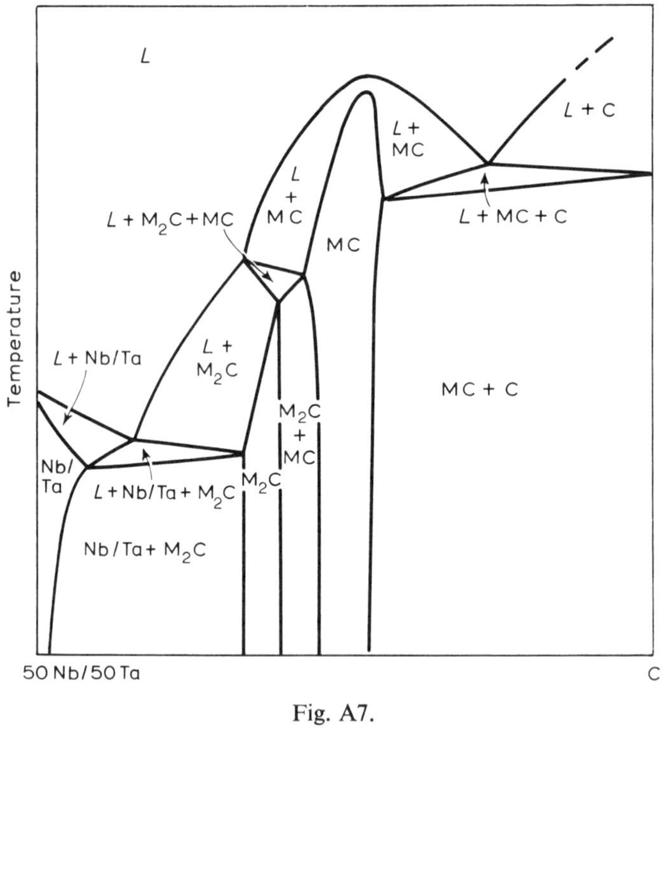

Fig. A7.

As–Ga–Zn

1. *Procedure*: From Fig. 7.2*d* the source composition at 1000°C lies in the liquid + GaAs region of the system. Draw a tie-line from GaAs through the source composition to intersect the 1000°C isotherm shown on the liquidus projection; this intersection point represents the composition of the liquid in equilibrium with GaAs.

Answer: ~51Ga 21As 28Zn in at%.

2. (a) See Fig. A8. Note: the solid solubility limits of As, GaAs, Zn_3As_2 and $ZnAs_2$ are very small; the lines joining respectively As–GaAs, As–$ZnAs_2$, GaAs–$ZnAs_2$, GaAs–Zn_3As_2 and Zn_3As_2–$ZnAs_2$ represent very narrow two-phase regions.

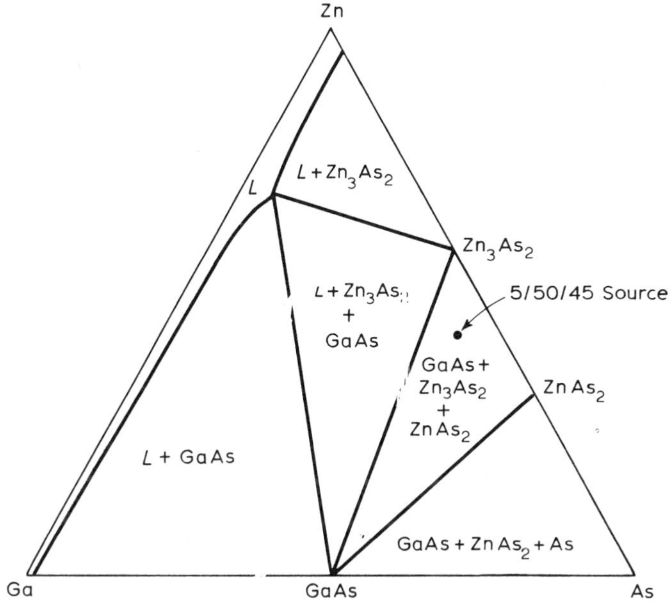

Fig. A8. Isothermal section of the Ga–As–Zn system at 700°C

(b) *Procedure*: Convert the composition of the alloy and of the relevant solid phases (i.e. GaAs, Zn_3As_2, $ZnAs_2$) from atomic to weight percentages; plot the resultant data as a tie-triangle and use the 'centre of gravity' principle.

Answer: ~11 GaAs, 54 Zn_3As_2 35 $ZnAs_2$ in wt%.

3. These sections are quasi-binary in nature and contain the equilibria $L \leftrightharpoons$ GaAs + Zn_3As_2 and $L \leftrightharpoons$ GaAs + $ZnAs_2$ respectively.

Au–Pb–Sn

1. $L \longrightarrow AuSn_2$; $L \longrightarrow AuSn_2 + Pb$; $L + AuSn_2 \longrightarrow Pb + AuSn_4$; $L \longrightarrow AuSn_4 + Pb$; $L \longrightarrow AuSn_4 + Pb + Sn$. ~40% Pb, 26% Sn, 34% $AuSn_4$.

2. *Procedure*: In Fig. 7.3*b* draw a line from Au to the solder composition; the amount of Au that can dissolve at 280°C is given by the composition corresponding to the intersection of this line with the 280°C isotherm.

Answer: ~26%Au.

3. Using the procedure indicated from question 2 above, the 40Pb–60Sn solder is found to dissolve ~10% Au at 220°C and the 50Pb–50Sn solder ~12% Au at 240°C. Considering the solidification sequences of the liquids formed by solution of Au for these two cases respectively, both will deposit primary $AuSn_2$, and the liquid compositions will then reach P_2 via the reaction $L \longrightarrow Pb + AuSn_2$. At P_2, $L + AuSn_2 \longrightarrow Pb + AuSn_4$ and under equilibrium $AuSn_2$ is consumed, since the overall compositions lie in the $L + Pb + AuSn_4$ region. Applying the centre of gravity construction the proportion of liquid (composition P_2) is ~80% for the 40Pb–60Sn solder as compared with ~45% for the 50Pb–50Sn solder. Solidification is completed by the sequence $L \longrightarrow Pb + AuSn_4$; $L \longrightarrow Pb + Sn + AuSn_4$ (at E). Application of the centre of gravity principle to the Pb–Sn–$AuSn_4$ triangle shows that ~34% $AuSn_4$ is formed in the 40Pb–60Sn solder as compared with ~37% $AuSn_4$ in the 50Pb–50Sn solder.

Al–Cr–Ni

1. (a) Consider alloy X (Fig. 7.4*d*): $L \longrightarrow$ primary β; the liquid composition moves on to the $L + \beta \longrightarrow \gamma'$ peritectic curve; as the peritectic reaction proceeds the alloy composition lies in the three-phase $L + \beta + \gamma'$ region; at 1340°C the liquid composition reaches L_p and the invariant reaction occurs $L_p + \gamma'_p \longrightarrow \beta_p + \gamma_p$; the liquid is consumed since X lies in the $\beta_p + \gamma_p + \gamma'_p$ tie-triangle and solidification is thus completed,

(b) Alloy containing 75 at% Ni, 22 at% Al, 3 at% Cr (represented schematically by *Y* in Fig. 7.4*d*); $L \longrightarrow$ primary γ; the liquid composition moves on to the $L \longrightarrow \gamma + \gamma'$ eutectic curve; solidification is completed to form $\gamma + \gamma'$.

2. The transition from β–γ to α–γ' equilibrium at 1000°C (a) Above invariant reaction temperature, (b) at invariant reaction, and (c) below invariant reaction temperature. (See Fig. A9.).

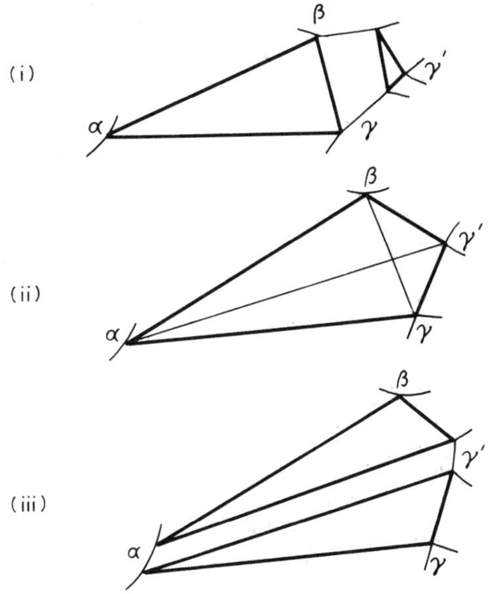

Fig. A9.

3. Figs. 7.4*e* and *f* show that the solubility of Al in γ decreases with increase in Cr content; similarly the $\gamma'/\gamma + \gamma'$ boundary moves to lower Al content with Cr additions. Considering a series of alloys with 15 at% Al and increasing Cr content up to 20%, at 750°C (Fig. 7.4*f*) the proportion of γ' increases, as indicated by the tie-line data. The lattice parameter mismatch between γ and γ' is small, and varies with composition, becoming zero for certain values of γ and γ' compositions.

Considering a series of alloys with 75 at% Ni (Fig. 7.4*g*), the $\gamma/\gamma + \gamma'$ transus temperature decreases with increase in Cr content; also in the $\gamma + \gamma'$ region at a given ageing temperature (e.g. 750°C), the proportion of γ' decreases (note that the γ/γ' tie-lines lie quite close to the plane of this vertical section at 750°C).

C–Cr–Fe

1. Ternary invariant reactions:

1585°C: $L + M_3C_2 \longrightarrow$ graphite $+ M_7C_3$
1415°C: $L + M_{23}C_6 \longrightarrow M_7C_3 + \alpha$
1275°C: $L + \alpha \longrightarrow \gamma + M_7C_3$
1230°C: $L +$ graphite $+ M_7C_3 \longrightarrow M_3C$ (see Fig. 4.13)
1160°C: $L + M_7C_3 \longrightarrow \gamma + M_3C$
1155°C: $L + M_3C \longrightarrow \gamma +$ graphite

2. $L \longrightarrow$ primary α (commencing at $\sim$1510°C); liquid composition changes until it reaches the $L + \alpha \longrightarrow \gamma$ curve; the liquid is consumed during this reaction and the resulting structure on cooling to just below the solidus ($\sim$1430°C) is $\alpha + \gamma$.

3. At room temperature, under equilibrium, the alloy consists of ferrite + $M_{23}C_6$. On heating to 1200°C a fully austenitic structure is produced; rapid quenching leads to martensite formation (together with a small amount of retained austenite); tempering at 700°C produces a structure of ferrite + $M_{23}C_6$. (The chromium content of the ferrite is reduced to less than 9% by the carbide precipitation and this adversely affects the corrosion resistance.) At 850°C the alloy composition lies in the $\gamma + M_{23}C_6$ phase region; the γ phase contains $\sim$9% Cr, 0·2% C, and rapid quenching produces martensite which on tempering transforms to ferrite + $M_{23}C_6$.

(Note: the lever rule cannot be applied directly to Figs 7.6*b* and *c* since triangular coordinates are not used and the percentage carbon scale is expanded relative to that for chromium.)

Cr–Fe–Ni

1. See Fig. A10.

2. $L \longrightarrow \alpha$ (commencing at $\sim$1430°C); liquid composition moves on to the $L \longrightarrow \alpha + \gamma$ reaction curve. The alloy composition lies close to the

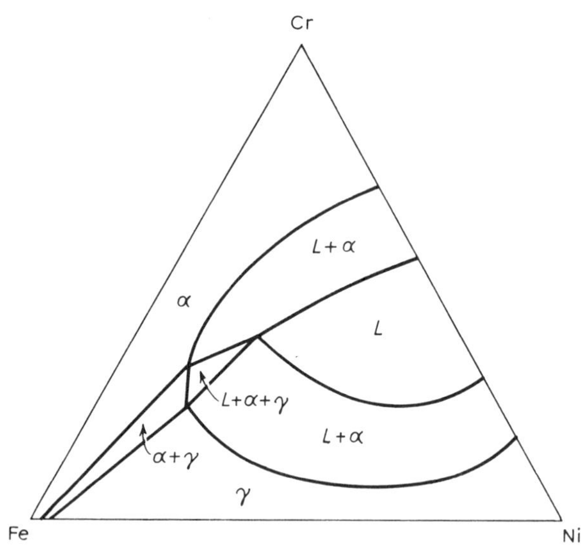

Fig. A10. Isothermal section of the Fe–Cr–Ni system at 1400°C

$\alpha + \gamma$ side of the $L + \alpha + \gamma$ tie-triangle in the 1400°C isothermal section. Thus the solidus for the alloy is ~1400°C; the application of the lever rule to the α–γ tie-line gives the proportions of α and γ as ~75% and 25%, respectively.

3. The Fe–18Cr–12Ni composition lies very close to the L, α, γ reaction curve; solidification commences at ~1470°C and proceeds via the $L \longrightarrow \alpha + \gamma$ reaction to produce an $\alpha + \gamma$ structure at ~1420°C. On cooling in the solid state the structure becomes single phase γ at a temperature not far below the solidus.[27] On further cooling, under equilibrium, σ phase begins to precipitate from γ at just above 650°C (Fig. 7.7*d*). If the alloy is rapidly cooled from 1000°C the austenite structure is retained to room temperature.

4. From Fig. 7.7*d*, along the line of constant Ni content (5%) alloys containing between ~21% and 30% Cr lie in the $\alpha + \gamma + \sigma$ tie-triangle and the proportion of σ increases from 0% to ~50% (calculated using the centre of gravity construction). From ~30% to 45% Cr in the $\gamma + \sigma$ region the proportion of σ increases to 100% (calculated from the lever rule and estimated positions of tie-lines). The single phase region is traversed between ~45% and 53% Cr and then as the Cr increases to ~63% (i.e. across the $\sigma + \alpha'$ region), the σ-phase content diminishes to zero.

Fe–O–S

1. See Fig. A11. Three tie-triangles (FeL_1L_2) are shown schematically in relation to the liquid miscibility gap to represent the progress of the monotectic reaction in alloy X. The three temperatures illustrated are indicated by subscripts *a*, *b*, *c* (see also reference 4).

.... represents the beginning of the monotectic reaction (X lies on the Fe L_{1a} side of the Fe L_{1a} L_{2a} tie-triangle)

------- at a lower temperature X lies within the FeL_{1b} L_{2b} tie-triangle.

— — — the end of the reaction is reached when X lies on the FeL_{2c} side of the FeL_{1c} L_{2c} tie-triangle.

2. For alloys that undergo the monotectic reaction the weight fraction of inclusions formed increases as the oxygen and sulphur contents of the alloy increase. However, the distribution of the compositions of the inclusions depends only on the oxygen/sulphur ratio. For alloys of high oxygen/sulphur ratio the bulk of the inclusions are rich in oxygen, while for low oxygen/sulphur ratios the majority of the inclusions are sulphur rich.

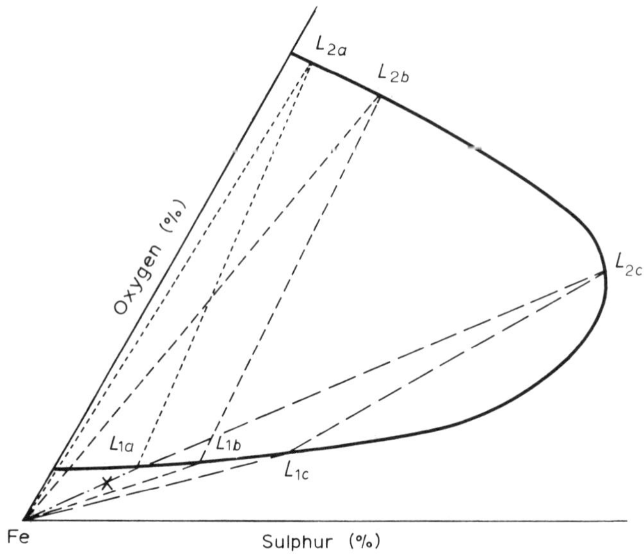

Fig. A11. Schematic representation of part of the liquidus projection of the Fe–O–S system

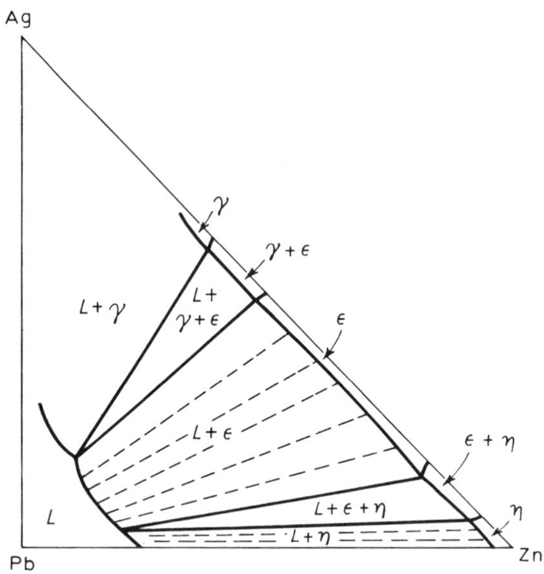

Fig. A12. Schematic isothermal section of Ag–Pb–Zn system at ~ 330°C

Ag–Pb–Zn

1. See Fig. A12.

2. The eutectic reaction $L \longrightarrow$ Pb + η runs from the temperature of the invariant down to the Pb–Zn binary eutectic at 318°C (Fig. 7.9*e*). Thus liquid solidifying along this curve is progressively depleted in silver content. By repeated zinc additions, the silver content of a bullion can be reduced to any desired figure as the Pb–Zn eutectic is approached. If the ternary invariant involving L, Pb, ε and η were a eutectic then the ternary eutectic composition would represent the lowest silver content attainable in lead by the use of zinc additions.

Al_2O_3–MgO–SiO_2

1. The section can be drawn from the information shown in Fig. 7.10*d* (see Fig. A13*a*).

2. (a) $L \longrightarrow$ spinel (Al_2O_3-rich); $L \longrightarrow$ spinel + corundum; L + corundum $\longrightarrow$ spinel + mullite at 1578°C; the overall composition lies just within the L + spinel + mullite region at 1578°C, thus corundum is consumed; $L \longrightarrow$ spinel + mullite.

(b) $L \longrightarrow$ spinel (Al_2O_3-rich); subsequent liquid path depends in detail on the spinel composition but liquid undergoes the L + spinel + mullite $\longrightarrow$ sapphirine reaction at 1482°C; mullite is consumed and the remaining liquid solidifies along the L + spinel $\longrightarrow$ sapphirine curve running down from 1482°C.

(c) $L \longrightarrow$ spinel (Al_2O_3-rich); $L \longrightarrow$ spinel + corundum at 1578°C; L + corundum $\longrightarrow$ spinel + mullite; corundum is consumed; $L \longrightarrow$ spinel + mullite; at 1482°C, L + spinel + mullite $\longrightarrow$ sapphirine; L, spinel and mullite consumed, leaving only sapphirine.

(d) $L \longrightarrow$ cristobalite/tridymite; $L \longrightarrow$ tridymite + cordierite.

(Note: when spinel is the primary phase, its precise composition is not known because of the wide solid solubility range and hence the path followed by the liquid composition over the primary liquidus surface is not known.)

3. See procedure discussed on page 48, Fig. 4.5.

(a) Melting commences at the 1365°C eutectic (forsterite + protoenstatite + cordierite $\longrightarrow$ liquid). The ceramic composition lies on the tie-line joining forsterite to the ternary eutectic composition. Application of the lever rule to this tie-line shows that ~20% liquid forms by melting of the ternary eutectic. On continued heating the proportions of liquid present at specific temperatures may be calculated by application of the

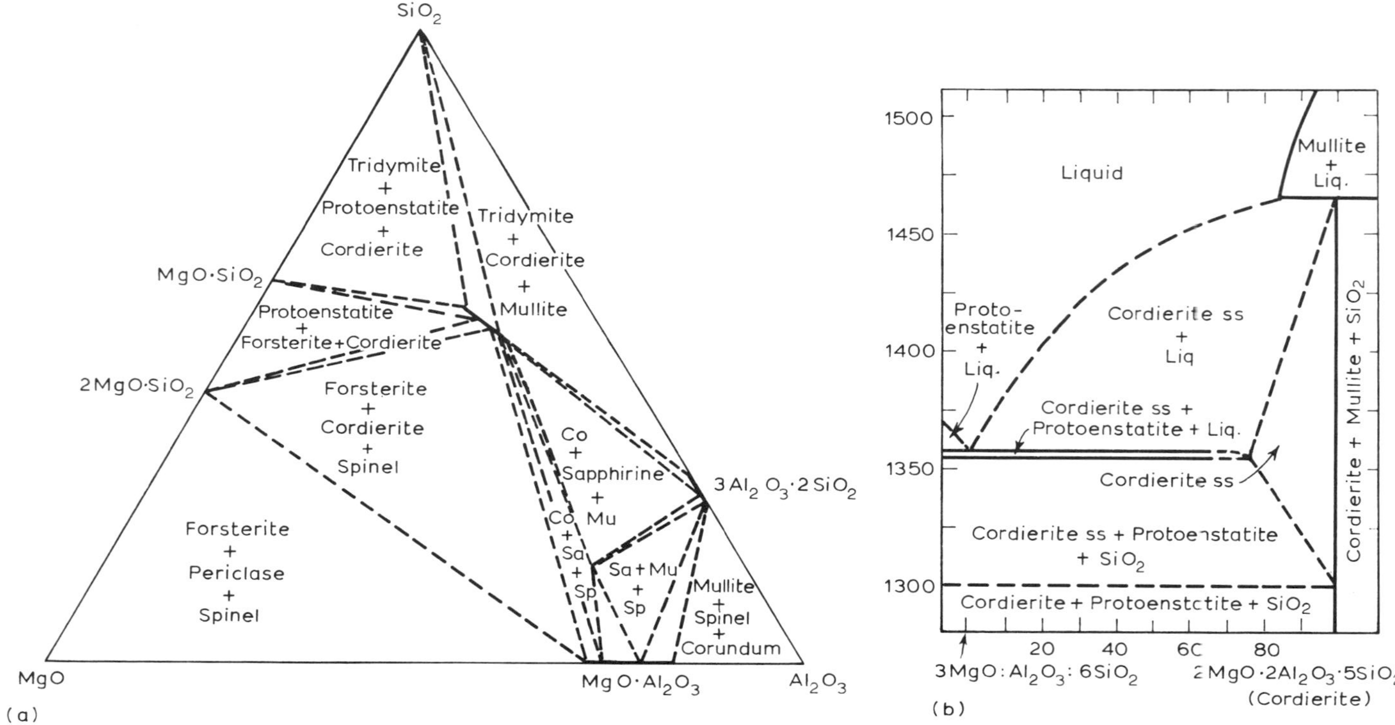

Fig. A13. Al_2O_3–MgO–SiO_2 system.
(*a*) Isothermal section representing the solid state constitution (at a temperature of ~ 1350°C) derived from the data in Fig. 7.10*d*.[37]
(*b*) Vertical section between cordierite and the composition $3MgO \cdot Al_2O_3 \cdot 6SiO_2$ (which lies close to the composition of the eutectic: liquid ⇋ tridymite + cordierite + protoenstatite).[36]
(*b* reproduced by permission of Oxford University Press from reference 36)

lever rule to the forsterite–liquid tie-lines (the liquid compositions being located at points on the liquidus isotherms). Thus ~30% liquid is present at 1600°C and ~39% at 1700°C; melting is completed at ~1850°C.

(b) Melting commences at the 1355°C eutectic (protoenstatite + tridymite + cordierite ⟶ liquid). Application of the centre of gravity principle to the liquid + protoenstatite + tridymite tie-triangle shows that ~30% liquid forms. On continued heating, melting proceeds with the liquid composition changing along the liquid/protoenstatite/tridymite reaction curve. (The ceramic composition lies on this curve.) At 1400°C ~35% liquid is present, and at ~1500°C, melting is completed.

(c) Melting commences at the 1355°C eutectic and ~40% liquid forms, assuming the maximum solubility range for cordierite shown in Figs 7.10*d* and A13*a*. However, it should be noted that the range is not known accurately, and a smaller range is shown in the vertical section of Fig. A13*b*[36], (the alloy composition lies close to this vertical section). Melting is complete at ~1430°C.

The variation of the proportion of liquid with temperature is important in the firing behaviour of ceramic compositions where the formation of ~20–35% of viscous liquid phase acts as a bond and aids densification. In the forsterite composition, (a), the amount of liquid does not increase rapidly with temperature (i.e. steep liquidus) and a broad firing range is therefore available. The shallower liquidus profiles of the steatite and cordierite type materials, (b) and (c), are not so favourable, but the addition of other materials may be used to improve the characteristics. It should be noted that in practice the processing of these ceramics from this system involves the use of clay and talc as raw materials (for a discussion of associated structural changes and of the kinetics of firing of ceramics see reference 32). It should also be borne in mind that the liquid phase commonly forms a glass on cooling instead of crystallizing.

4. The 42% Al_2O_3–58% SiO_2 material is solid at 1460°C but with 1% MgO the composition lies in the liquid (g) + mullite + cristobalite region (Fig. 7.10*e*). Applying the centre of gravity principle to the tie-triangle the percentage liquid is ~10%.

References

Selected references sources are quoted [13–38] for the phase diagrams presented in Chapter 7. It should be noted that it is not uncommon to find differences between the results reported by various authors for a given system; these differences may be significant in relation to phase compositions and the temperatures and natures of the reactions.

[1] J. W. Gibbs, *The Collected Works of J. Willard Gibbs*, v. I. pp. 54–371, Longmans, Green and Co., New York, 1928.

[2] R. Vogel, *Die Heterogenen Gleichgewichte*, Akademische Verlagsgesellschaft, Geest Portig, Leipzig, 1959.

[3] J. E. Ricci, *The Phase Rule and Heterogeneous Equilibrium*, D. van Nostrand Co. Inc., New York, 1951.

[4] G. Masing (translated by B. A. Rogers), *Ternary Systems*, Reinhold Publishing Corporation, New York, 1944.

[5] J. S. Marsh, *Principles of Phase Diagrams*, McGraw Hill Book Co., New York, 1944

[6] W. Hume-Rothery, J. W. Christian, and W. B. Pearson, *Metallurgical Equilibrium Diagrams*, The Institute of Physics, London, 1952.

[7] F. N. Rhines, *Phase Diagrams in Metallurgy*, McGraw Hill Book Co., New York, 1956

[8] E. M. Levin, H. F. McMurdie, and F. P. Hall, *Phase Diagrams for Ceramists*, The American Ceramic Society Inc., 1956.

[9] Metals Handbook, *American Society for Metals*, 1948.

[10] M. Hillert, *J. Iron and Steel Institute*, **189,** 224, 1958.

[11] H. Lipson and A. J. C. Wilson, *J. Iron and Steel Institute*, **142,** 107, 1940.

[12] H. W. L. Phillips, *Equilibrium Diagrams of Aluminium Alloy Systems*, The Aluminium Development Association, London, 1961.

[13] L. E. Toth, *Transition Metal Carbides*: *Refractory Materials Series*, **7,** Academic Press, New York, 1971 (after E. Rudy, *Compendium of Phase Diagram Data*, AFML-TR-65-2, part V, 1960).

[14] M. B. Panish, *J. Phys. Chem. Solids*, **27,** 291, 1966.

[15] H. C. Casey Jr and M. B. Panish, *Met. Trans. A.I.M.E.*, **242,** 406, 1968.

[16] K. K. Shih, J. W. Allen and G. L. Pearson, *J. Phys. Chem. Solids*, **29,** 379, 1968.

[17] M. B. Panish, *Phase Diagrams*, vol 3, Academic Press, New York, 1970.

[18] A. Prince, *J. Less Common Metals*, **12,** 107, 1967.

[19] A. Taylor and R. W. Floyd, *J. Inst. Metals*, **81,** 451, 1952–53.
[20] A. Taylor, *Trans. A.I.M.M.E.* **206,** 1356, 1956.
[21] C. T. Sims and W. C. Hagel, *The Superalloys*, John Wiley and Sons, New York, 1972.
[22] A. Prince, *Alloy Phase Equilibria*, Elsevier, Amsterdam, London, New York, 1966.
[23] W. D. Forgeng and W. D. Forgeng, Jr. in *Metals Handbook*, eighth edition, **8,** Metallography, Structure and Phase Diagrams, American Society for Metals, Metals Park, Ohio, 1973.
[24] K. Bungardt, E. Kunze and E. Horn, *Archiv. f. das Eisenhüttenwesen*, **3,** 193, 1958.
[25] L. Colombier and J. Hochmann, *Stainless and Heat Resisting Steels*, Edward Arnold Ltd, London, 1967.
[26] Von W. Tofaute, C. Küttner and A. Büttinghaus, *Archiv. f. das Eisenhüttenwesen*, **9,** 607, 1935–36.
[27] V. G. Rivlin and G. V. Raynor, *International Metals Reviews*, **21,** no. 1, 1980.
[28] F. H. Keating, *Chromium–nickel Austenitic Steels*, Butterworths, London, 1956.
[29] J. C. Yarwood, M. C. Flemings and J. F. Elliott, *Met. Trans.* **2,** 2573, 1971.
[30] T. R. A. Davey, *Trans. A.I.M.M.E.*, **200,** 838, 1954.
[31] Ernest M. Levin, Carl R. Robbins and Howard F. McMurdie, *Phase Diagrams for Ceramists*, American Ceramic Society Inc., Columbus, Ohio, 1964 (Supplement 1969).
[32] W. D. Kingery, H. K. Bowen and D. R. Uhlmann, *Introduction to Ceramics*, Second Edition, John Wiley and Sons, New York, London, 1976.
[33] R. Kiessling and N. Lange, *Non-metallic inclusions in steels*, Part II, Iron and Steel Institute, London, 1966.
[34] A. Muan and E. F. Osborn, *Phase Equilibria Among Oxides in Steel Making*, Addison-Wesley Publishing Co. Inc., Reading, Massachusetts, 1965.
[35] E. G. Ehlers, *The Interpretation of Geological Phase Diagrams*, W. H. Freeman and Co., San Francisco, 1972.
[36] W. Schreyer and J. F. Schairer, *J. Petrology*, **2,** 3, 324, 1961.
[37] E. F. Osborn and Arnulf Muan, *Phase Equilibrium Diagrams of Oxide Systems*, Plate 3, The American Ceramic Society and the Edward Orton Jr Ceramic Foundation, 1960.
[38] F. Tamás and I. Pál, *Phase Equilibria Spatial Diagrams*, Iliffe Books (Butterworth and Co.), London, 1970.

Other sources

J. L. Haughton and A. Prince, *The Constitutional Diagrams of Alloys*, Second edition, The Institute of Metals, London, 1956.

A. Prince, *Multicomponent Alloy Constitution Bibliography 1955–73*, The Metals Society, London, 1978.

Phase diagrams for binary systems may be found in sources such as references 23 and 31 and in the following:

M. Hansen and K. Anderko, *Constitution of Binary Alloys*, McGraw Hill, New York, 1958; R. P. Elliott, First supplement, 1965: F. A. Shunk, Second supplement, 1969.

Index

Graph Paper for Calculations

This page can be photocopied to provide graph paper suitable for solution of the problems and for other rough calculations.

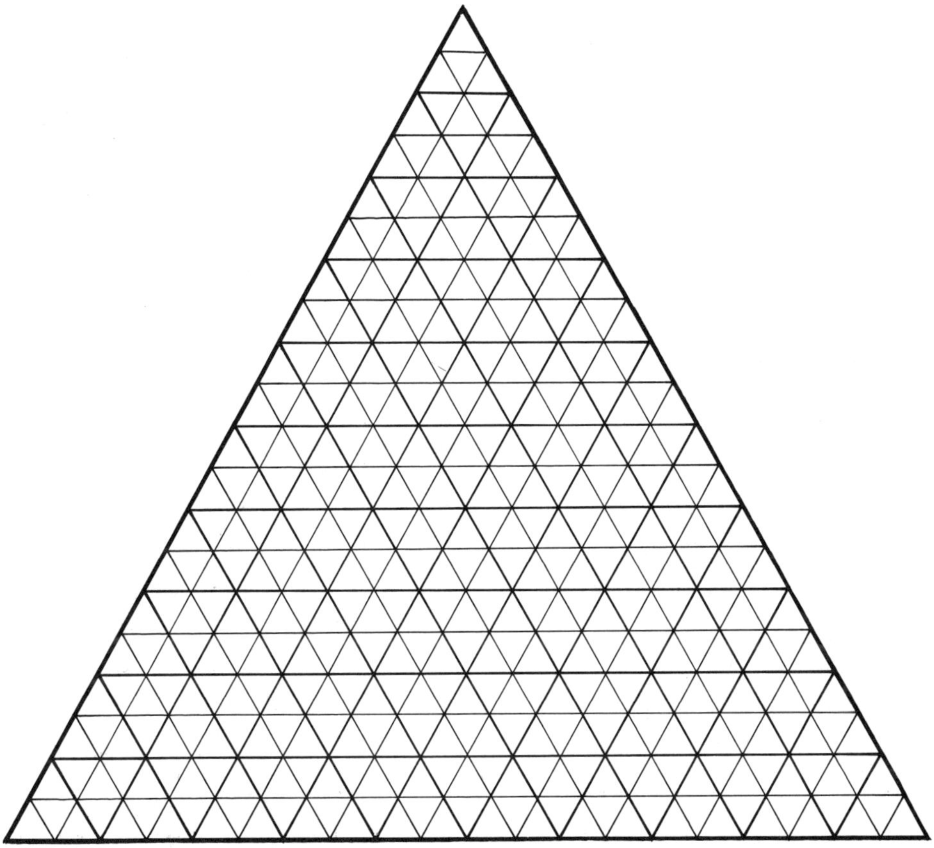